T0295302

From the Atom to Living Systems

From the Atom to Living Systems

Systems

A Chemical and Philosophical Journey into Modern and Contemporary Science

MARINA PAOLA BANCHETTI-ROBINO AND
GIOVANNI VILLANI

OXFORD
UNIVERSITY PRESS

OXFORD
UNIVERSITY PRESS

Oxford University Press is a department of the University of Oxford. It furthers
the University's objective of excellence in research, scholarship, and education
by publishing worldwide. Oxford is a registered trade mark of Oxford University
Press in the UK and certain other countries.

Published in the United States of America by Oxford University Press
198 Madison Avenue, New York, NY 10016, United States of America.

CIP data is on file at the Library of Congress

ISBN 978-0-19-759890-0

DOI: 10.1093/oso/9780197598900.001.0001

Printed by Integrated Books International, United States of America

Table of Contents

Introduction: A Systemic Chemistry

This book represents a philosophical and scientific journey that will proceed according to three stages of increasing levels of complexity, beginning with the atom, then moving to molecules and macromolecules, and finally moving to the threshold of life. In the first stage, we will analyze the transition from the qualitatively undifferentiated atom to differentiated atoms, that is, to atoms of iron, of copper, of hydrogen, and the like. In the second stage, we will analyze the passage from atoms to molecule, and in the third stage, we will examine the passage from molecule to macromolecule. Simultaneously, as we pass through these levels of complexity, we will also examine the transformation of the fundamental philosophical concept of interaction between entities. Thus, in the first stage, we transition from mechanical interaction to interaction involving chemical affinity. In the second stage, we transition from affinity to molecular structure, and finally, in the third stage, we transition from molecular structure to the form and superstructure of macromolecules. To introduce such a wide-ranging exploration, we consider it appropriate to address the philosophical perspective regarding nature that informs this book, which can best be characterized as the systemic perspective. Since the term *systemic* has acquired a wide and diverse range of meanings, we will clarify our approach by specifying what we mean by the term *system.*

In order to frame this book within a more general metascientific context, we acknowledge that this work is positioned within the still emerging but rapidly expanding perspective that seeks to overcome a strictly reductionist scientific vision and to replace it with a systemic approach. Epistemological analyses of science often consider only two possible, albeit opposing, philosophical positions, that is, the reductionist view and the holistic view. The systemic approach that we defend in this book represents a type of intermediate position between reductionism and holism. In fact, the concept of system merges reductionism and holism by considering the essential function of both the parts and the whole, as well as the relationships within and between these. The French philosopher and sociologist Edgar Morin[1] states that the concept of system/organism cannot be resolved in the relationship between the parts. As this book argues, since the reductionist explanation focuses only on the parts, it cannot be considered

[1] Morin, *Il metodo 1. La natura della natura.*

From the Atom to Living Systems. Marina Paola Banchetti-Robino and Giovanni Villani, Oxford University Press.

sufficient in the study of systems/organisms because it eliminates, from the very start, the complex unit that supports the system. However, the holistic approach that is opposed to reductionism also cannot be considered totally effective for highlighting systemic complexity. This is because this approach endeavors to reduce all systemic concepts to the concept of totality. Both reductionistic and holistic explanations suffer from being far too simplistic. Both of these perspectives spring from the scientific and philosophical predilection for simplicity, so that a simple rationalization must ultimately explain the complexity of natural processes.

This book accepts the importance of describing systems in a holistic manner. In fact, many properties of systems can be understood through their levels of complexity, through horizontal explanations that require a single plane of complexity. We also acknowledge the importance of approaching some explanations in a reductionist manner, beginning with the decomposition of an entity or process into its constituent parts and then studying their interactions, seeking to reconstruct the global properties of the entity or process according to its underlying levels. This requires a *bottom-up* explanation because it provides a lower-level explanation for higher-level entities or processes. This type of explanation, which is extensively used in many fields, allows us to successfully understand many important aspects of reality, as the widely reductionist history of modern science demonstrates. Therefore, the merits of both types of analysis, in terms of levels of complexity and in terms of constituents, cannot be denied. However, neither of these opposing approaches can yield individually a complete and balanced description of reality.

In this book, we contrast these two philosophical and scientific positions, that is, the purely holistic and the purely reductionistic perspective. We also challenge the idea that either of these two approaches will suffice, in itself, to yield a systematic and complete understanding of nature.

The first holistic perspective was adopted by Ludwig von Bertalanffy, who famously stated "General system theory . . . is a general science of 'wholeness' which up till now was considered a vague, hazy, and semi-metaphysical concept."[2] The second and reductionist perspective, on the other hand, has predominated modern science since the 17th century and has given rise to two significant problems: (1) The scientific expectation that a Theory of Everything (TOE) would be sufficient to explain everything, from elementary particles to life itself, and could even be extended to those phenomena with which the social sciences and the humanities are concerned and (2) the metascientific predilection for reducing social and collective behavior to individual psychology,

---

[2] Bertalanffy, *General System Theory*, p. 37.

psychology to applied biology, biology to applied chemistry, and chemistry to applied physics.

In this book, we demonstrate that the above type of reductionism is highly problematic since chemistry, for example, is central for understanding not only the phenomena studied in biology (in a bottom-up fashion) but also the entities and processes studied in physics (in a top-down fashion). Thus, in spite of the still-prevailing view that chemistry can, at least in principle, be reduced to physics, one of the central aims of this book is to demonstrate that chemistry is completely irreducible to physics. In addition, although chemistry is crucial in understanding biological and other higher-level phenomena, we will argue that the humanities and social sciences cannot be reduced to psychology, psychology cannot be reduced to biology, and biology cannot be reduced to chemistry. In spite of the still-dominant nonliving/living and animate/human dichotomies, our aim is to avoid such dichotomies and instead focus on what unites the different levels of reality and what allows science to comprehend the way in which complex systems function.

In order to do so, we must clarify how we will use the terms *system* and *organism*, since these terms have been given multiple definitions throughout the history of modern science and it is not the task of this book to analyze and compare these various meanings. Thus, we propose to use the terms *system* and *organism* interchangeably and to indicate exclusively a structured or organized set of interrelated elements, whose organization configures the interrelations between the parts of the system, the relations between the parts and the whole, the relations between the whole and its parts, and the relations between the system and its environment. In fact, it is precisely organization and interrelations that transform a set of discrete entities into a system, so that leaving out the concept of organization amounts to conceptualizing systems without solid foundations. Von Bertalanffy himself stated that a system is a set of interrelated unities and it is organization that forms, in space and time, a new reality that we call a "system." In cyclical fashion, organization produces the order that maintains the organization that produced it and we will return later to this cyclical process.

The concept of a system as a structured and/or organized whole has long been underestimated in the natural sciences, while it has always been of central concern to the humanities and the social sciences. In fact, one of the causes for the long-standing tension between the natural sciences and the social sciences has been precisely this tacit, and at times explicit, disagreement over the ontological status and epistemological value of the concept of system. Thus, in order to remove the barriers between the natural sciences, the social sciences, and the humanities, it is crucial to recognize the importance of a systemic approach to the study of nature at all its levels.

The whole and its parts are the necessary levels for the constitution of an organized system, and organization is itself a global property of the system. In the context of chemistry, the organization of a molecule is specified by the global concept of molecular structure, which, as shall be demonstrated, cannot in any case be reduced to the "spatial arrangement of the constituent atoms," as is sometimes mistakenly asserted in chemistry textbooks. In fact, there are three key assertions that articulate the idea of a system, assertions that may appear contradictory but that are mutually necessary and complementary for defining the concept of a system.

First, the system must be regarded as more than the summation of its parts either considered individually or juxtaposed. The system's organization and global unity must also be considered, as well as the new qualities and properties that emerge precisely from its organization and global unity and that differentiate it from a mere aggregate of parts. However, such changes occur not only at the level of the system as a whole but also at the level of the individual parts. Remarkably, the feedback loop of macro-emergence on the parts of the system produces instances of micro-emergence. One example of macro-emergence is that of the atom, which is a system that has its specific properties such as stability, a property that the constituent particles do not display when considered in isolation. Other examples include matter itself, life, consciousness, and linguistic meaning, which can all be understood as emergent qualities of highly complex systems.

Not only is the whole more than the summation of its parts. Each part of the whole is more than a mere part, precisely as a consequence of being a constituent of a whole. Therefore, the concept of emergence requires a change in the manner through which science explains systems, in order to incorporate the relationship between the individual constituent units and the complex whole. This new way of understanding complex systems also requires transcending the simple hierarchy between infrastructure and superstructure in favor of a conception of organization in which the whole that is produced retroactively transforms the parts that have produced it. The link between formation and transformation is a key systemic principle: *Everything that forms, transforms.*

Second, the system is less than the sum of its parts either considered individually or juxtaposed. While it has often been noted that the whole is more than the summation of its parts, the opposite notion has rarely been recognized. However, the idea that the system is less than the sum of its parts is deducible from the concept of organization itself and is actually easier to understand than the concept of emergence. The presence of organization is equivalent to the existence of constraints upon the parts, restrictions on the states that the constituents of the system can assume. Each systemic association entails constraints, such as reciprocal constraints between the parts, constraints of the parts upon the whole, constraints of the whole upon the parts. Thus, in every system, one must evaluate not only the qualities gained through emergence but also the losses due to the constraints. The acquisition and the loss of qualities are indications that the individual constituents that become part of a system are, first and foremost, transformed into parts of a whole. This is another key systemic principle: *Everything that forms is transformed.*

This occurs, for example, in the cells of an organism. It is well known that such cells carry the total genetic information of that organism. However, most of this information is suppressed, and only the small amount of information that corresponds to the specialized activity of the cell can be expressed. This is even more evident in the molecule, which is the focus of this book. In this case, it is only the constraints themselves that prevent atoms from moving away from each other and creating a new system. This analysis can, of course, be carried to even the highest levels of organization, such as social systems, in which the constraints are even more obvious.

Third, the system is an intrinsically dynamic entity. In fact, dynamism is another fundamental feature of the concept of system, which describes its dual nature as both entity and process. Therefore, the classic philosophical and scientific distinction between entity and process must be revised within the systemic perspective, since both of these aspects converge in a system. In the identity closure of a system, which separates it conceptually and physically from its environment, some dynamic processes represent the internal dynamism of that system, while others constitute the dynamic interaction of the system as a whole with the outside. A system is, therefore, an entity with intrinsic dynamic interactions between its constituents and in global dynamic interaction with its environment. Since these two sets of processes often occur at different times, the separation and modeling between the system and its environment are simplified.

There are, however, situations in which these processes are entangled and separation is difficult. A good example of this situation is the coupling between internal and external processes in living beings, and an even better example is that of human problems, in which the internal and external processes are often entangled. Edgar Morin has discussed such active systems, and we take his lead in employing this concept. Since many biologists focus on organisms considered as entities, many of them have concentrated on studying the constituents essential to life, while other biologists have privileged the process characteristics of organisms. From a systemic perspective on life, both the concept of entity and the concept of process are essential and must coexist in a balanced manner within the systemic concept of *dynamic entity*.

The unitary theory of organized sets was developed in the second half of the 20th century, and many authors have openly expressed the belief that there is a universal way of considering such composite entities. Regardless of the type of components and the respective relationships that "bind" them in a system, the formal correspondence between the general principles has led to the conception of a general theory of systems that is intended as a new scientific doctrine dealing with formal principles applicable to all systems in general. In spite of this theory, which considers systems in a manner similar to the physical sciences, our approach in this book is more closely in line with the general systemic approach that is often referred to as systemics.

The first steps toward the development of systemics were made in the behavioral sciences, in relation to the analysis of the internal processes of organisms, and in the study of social organizations whose parts were to be related to the whole within which they belonged. Early examples of this effort include the organismal biology of Jennings, Cannon, and Henderson, as well as the first Gestalt theory and its derivations. Many of these attempts were based on the closed system model, but an important step forward was later made when systems were finally seen as closely linked to their respective environments, thus leading to open system models. Von Bertalanffy was the first to stress the importance of the condition of either closure from or openness to the environment in living organisms. He advanced the concept of organism as an open system in 1932, followed by the development of the general principles of kinetics and their implications in the field of biology,[34] and there were later studies on the thermodynamics of open systems.[56] The difference between open and closed systems has been used to distinguish living beings from inanimate objects since, unlike inanimate material objects, all living organisms import certain types of materials from their external environment, transform them according to their needs and characteristics, and export other types of materials back into their environment. Von Bertalanffy soon discovered, however, that such open conditions can also be found in some inorganic systems.[789]

In this work, we are also indebted to Edgar Morin for laying the foundations of a truly modern philosophical approach to systems, based on the crucial notions of organization and complex organized unity, and for developing the ternary concept of interaction–organization–system in all of its fecundity. For Morin, a new approach to the system as a complex organized unity requires us to stop focusing on eliminating the paradox that underlies this idea and stop trying to break the circularity of relationships that are established in the process of describing the wholes and its parts, as well as the unity and the diversity within it. Instead, we should conceive of all these aspects together, both in their complementary and

[3] Bertalanffy, "Der Organismus als physikalisches System betrachtet."
[4] Bertalanffy, *Problems of Life.*
[5] Prigogine, *Étude Thermodynamique des Phenomènes Irréversibles.*
[6] Prigogine, *Introduction to Thermodynamics of Irreversible Processes.*
[7] Bertalanffy, *Theoretische Biologie.*
[8] Bertalanffy, "Zu einer allgemeinen Systemlehre."
[9] Bertalanffy, *Das Biologische Weltbild.*

in their antagonistic relationships. A systemic approach, therefore, inevitably generates and requires a kind of circular explanation. This type of explanation was first generated within a cybernetic environment and, more specifically, from Norbert Wiener's idea of the corrective feedback loop that was initially used to study the coupling of a computer and a missile-guiding radar on a target at variable positions. This idea was then substantially broadened with the development of automatic adjustments in which negative feedback devices cancel out deviations, in relation to the norms assigned to the machines. In fact, the classical and simple conception of causality is also invested with the idea of both positive and negative feedback. The fact that an effect can change its cause, becoming the cause of this change, implies leaving the linear concept of cause/effect. This feedback creates a circular causality in which the cause creates an effect that modifies the cause.

Negative feedback, which cancels a deviation from a norm, is a process of nullifying the effects to the system produced by external causes. Thus, the idea of negative causality[10] arises wherever there is regulation. For example, lowering the temperature outside a living organism should result in lowering the internal temperature of such an organism. However, the internal temperature remains constant, but this is not due to insulation, which would make the body insensitive to external variations. It is due to the production of heat within the organism. Ambient cooling increases heat production in homeothermic animals, as it does in the boiler of a regulated thermal machine. Thus, this means that the external cooling causes an internal heating. Consequently, not only does the cause not bring about the expected effect, but it actually brings about an effect that is contrary to the one that would have been expected.

In this process, Morin distinguishes an external cause (*eso-causa*) from an internal cause (*endo-causa*), which are different by nature. The *eso-causa* is general and is not linked to the organization of the system, while the *endo-causa* is local and is intimately linked to the active organization of the system.[11] The classical conception of causality as external could only account for states of equilibrium and imbalance. However, stationary states are established with the process of circular causality (cause→effect→cause→ . . .) that, by nullifying, contrasting, and even reversing external causality ultimately protects and maintains internal homeostasis. Circular causality is, therefore, inherent in organized systems. We will later discuss how important this type of circular causality is for living organisms. At this point, it will suffice to cite the book *About Life* with regard to precisely this matter: "Cell structure and organisation are necessary for metabolism and

[10] Bateson, "Cybernetic Explanation."
[11] Morin, *Il metodo 1. La natura della natura.*

metabolism is necessary for cell structure and organisation. In our opinion, this absolute and intimate interdependence is part of the essence of 'livingness.'"[12]

The systemic approach teaches us that, when we study systems at different levels of complexity, we must not push the analysis to the first element, that is, to the lowest level of complexity. In fact, the more internal and external connections are developed in a system, the more complex are the circularities that are at the base of this association and the more this will depend on the system's lower levels of complexity as well as on its environment (a level of greater complexity). In other words, the more a system is integrated inwardly and outwardly, the deeper must the analysis be pushed in order to highlight the details required to understand the system as a whole.

This systemic approach contrasts with the dominant reductionist perspective according to which the lowest level of organization is that which contains the foundational elements that serve as the initial source of explanation. But, in fact, this is the case only up to a certain number of levels of complexity. Reductionism is defined as

> a set of ontological, epistemological, and methodological claims about the relations between different scientific domains. The basic question of reduction is whether the properties, concepts, explanations, or methods from one scientific domain (typically at higher levels of organization) can be deduced from or explained by the properties, concepts, explanations, or methods from another domain of science (typically one about lower levels of organization). Reduction is germane to a variety of issues in philosophy of science, including the structure of scientific theories, the relations between different scientific disciplines, the nature of explanation, the diversity of methodology, and the very idea of theoretical progress, as well as to numerous topics in metaphysics and philosophy of mind, such as emergence, mereology and supervenience.[13]

For the staunch reductionist, there is an explanation at every level that always faces in the direction of lower levels. Faced by criticism from many condensed matter physicists, the 1979 Nobel Prize-winning physicist Steven Weinberg clearly expressed this view when he requested appropriations to build a particle accelerator:

> Still, relying on this intuitive idea that different scientific generalizations explain others, we have a sense of direction in science. There are arrows of scientific explanation, which thread through the space of all scientific generalizations.

[12] Agutter and Wheatley, *About Life. Concepts in Modern Biology*, 36.
[13] Brigandt and Love, *"Reductionism in Biology," Stanford Encyclopedia of Philosophy*.

Having discovered many of these arrows, we can now look at the pattern that has emerged, and we notice a remarkable thing: perhaps the greatest scientific discovery of all. These arrows seem to converge to a common start anywhere in science and, like an unpleasant child, keep asking "Why?" you will eventually get down to the level of the very small. . . . I have remarked that the arrows of explanation seem to converge to a common source, and in our work on elementary particle physics we think we're approaching that source. There is one clue in today's elementary particle physics that we are not only at the deepest level we can get to right now, but that we are at a level which is in fact in absolute terms quite deep, perhaps close to the final source. . . . There is reason to believe that in elementary particle physics we are learning something about the logical structure of the universe at a very, very deep level. The reason I say this is that as we have been going to higher and higher energies and as we have been studying structures that are smaller and smaller we have found that the laws, the physical principles, that describe what we learn become simpler and simpler. . . . . What I am saying is that the rules that we have discovered become increasingly coherent and universal. We are beginning to suspect that this isn't an accident, that it isn't just an accident of the particular problems that we have chosen to study at this moment in the history of physics, but that there is simplicity, a beauty, that we are finding in the rules that govern matter that mirrors something that is built into the logical structure of the universe at a very deep level.[14]

It is fortunate, however, that not all physicists agree with Weinberg. One of these physicists is 1977 Nobel Prize winner Philip W. Anderson, who stated:

The arrogance of particle physicist and his intensive research may be behind us (the discoverer of positron [Paul Dirac] said "the rest is chemistry"), but we have yet to recover from that of some molecular biologists, who seem determined to try to reduce everything about the human organism to "only" chemistry, from the common cold and all mental disease to the religious instinct. Surely there are more levels of organization between human ethology and DNA than there are between DNA and electrodynamics, and each level can require a whole new conceptual structure.[15]

Anderson's beautiful 1972 paper "More Is Different," from which this citation is taken, became one of the manifestos of antireductionism. We note that Anderson's emphasis is not placed on the "system that is more than the sum of its constituents," but on the fact that the system generates difference and novelty.

[14] Weinberg, "Newtonianism, Reductionism and the Art of Congressional Testimony."
[15] Anderson, "More Is Different."

Anderson intuits this movement from quantity to quality, and this intuition is an essential characteristic of modern antireductionism.

It would not be out of place to recall that, in the 1987 paper cited above, Weinberg states:

> On the other hand, we also would say that chemical behaviour, the way molecules behave chemically, is explained by quantum mechanics and Coulomb's law, but we don't really deduce chemical behaviour for very complex molecules that way. We can for simple molecules; we can explain the way two hydrogen atoms interact to form a hydrogen molecule by solving Schrodinger's equation, and these methods can be extended to fairly large molecules, but we can't work out the chemical behaviour of DNA by solving Schrodinger's equation. In this case we can at least fall back on the remark that although we don't in fact calculate the chemical behaviour of such complicated molecules from quantum mechanics and Coulomb's law, we could if we wanted to. We have an algorithm, the variational principle, which is capable of allowing us to calculate anything in chemistry as long as we had a big enough computer and were willing to wait long enough.[16]

Although we have addressed reductionism as though it were a homogeneous perspective, we will actually critique several types of reductionism in this book. The first type of reductionism is exemplified by the Enlightenment scientist Pierre-Simon Laplace and will be referred to as Laplacian reductionism. Laplace was firmly convinced that "An Intelligence that knows, at a given moment, all the forces from which nature is animated and also the disposition of all the entities that compose it and that, moreover, is deep enough to submit these data to analysis, would embrace in the same formula the movements of the largest bodies of the universe and the lightest atoms; for such an Intelligence, nothing would be uncertain and it would be fully cognizant of both the past and the future."[17] According to Laplacian reductionism, every single complex event is explained by determining the characteristics of the initial link in the chain of events, that is, the level of elementary particles and the laws that govern their interactions. Once the level of elementary particles and the "Law of Everything" are known, Laplace believes that every event at every level is implicitly contained, at least conceptually, in the initial level. His perspective forms the basis for Weinberg's reductionist standpoint.

The second type of reductionism is hierarchical reductionism, and, according to this view, a complex event is determined by a whole chain of explanations, link

[16] Weinberg, "Newtonianism, Reductionism and the Art of Congressional Testimony," 435.
[17] Laplace, "*Essai philosophique sur les probabilités,*" 4.

after link. The English biologist Richard Dawkins is a famous contemporary defender of this type of reductionism. He states:

> The hierarchical reductionist, on the other hand, explains a complex entity at any particular level in the hierarchy of organization, in terms of entities only one level down the hierarchy; entities which, themselves, are likely to be complex enough to need further reducing to their own component parts; and so on. . . . This was the point of explaining cars in terms of carburettors rather than quarks. But the hierarchical reductionist believes that carburettors are explained in terms of smaller units. . . , which are explained in terms of smaller units. . . which are ultimately explained in terms of the smallest of fundamental particles. Reductionism, in this sense, is just another name for an honest desire to understand how things work.[18]

Both the Laplacian and the hierarchical variants of reductionism, as well as other variants that are not being considered here for the sake of brevity, imply that explanations must move only in one direction, that is, bottom-up. However, cyclical explanations such as those in the "systemics" perspective that we endorse always involve both bottom-up and top-down explanations, in addition to their explanatory closure within a cycle. Thus, at one level of explanation, a human being may be nothing more than a set of 10^{27} protons, neutron, and electrons. But at the level of exhibited behavior, a human being is more than the mere cluster of its constituent particles, more than the summation of its parts and their interactions. It would be mistaken to say that the human being is completely understood, without remainder, once all of the relations between particles are explicitly and fully considered. Those who endorse this flawed perspective do not fully understand what it would mean to explicitly and fully consider the relations between 10^{27} protons, neutron, and electrons, particularly since these particles are not themselves elementary and to stop at such particles would not be consistent with a true reductionism. Additionally, defenders of this approach are also clearly unaware of the distinction between an isolated entity and an entity that functions as a component within a system.

In the debate between reductionism and antireductionism, it is essential to clarify what we mean by explanation. To say that the only type of explanation is one that identifies the final cause of all higher-level events within the relations between constituent particles is misleading. For example, the properties of human beings are manifestations of the particular level and type of complexity that is being considered. Responding to complex events by taking for granted

---

[18] Dawkins, *The Blind Watchmaker: Why the Evidence of Evolution Reveals a Universe without Design*, 13.

that only reductive mechanistic descriptions can "explain" such events is precisely to confuse description with explanation. Such descriptions are precisely only that, descriptions, but they are not explanations. If reductionism advocates that all explanations of complexity must be sought at the lower level of the most elementary constituents of matter, then reductionism is completely false both in principle and from a practical standpoint. Conversely, we will see that new types of organization occur at each level of complexity, as we proceed from quarks to nucleons, to atoms, to molecules, to cells, and so forth. Each of these new entities will display specific behaviors, that is, manifestations unique to that particular level of organization as achieved under specific environmental conditions.

As has already been stressed, this book defends a clearly antireductionist approach. However, we wish to clarify an important point regarding antireductionism in biology and, more specifically, in the philosophy of biology because we believe that the basic thesis of many antireductionist positions in biology is mistaken and may actually harm the cause of antireductionism. The basic thesis of antireductionism is, in fact, the antinomy between the living and the nonliving. It is no coincidence that antireductionists place emphasis on the insufficiency of the physical and chemical approaches for understanding living organisms. However, we regard the solution proposed by such antireductionists as being worse than the problem they are trying to resolve. If one opposes reductionism by presenting an absolute separation, a philosophical antinomy, between the living and the nonliving, then one is creating a double space of reduction. If one adds to this a dichotomy of the animate and the human, the problem is further complicated. This becomes evident when we examine the question of the origin of life and the origin of the human species. Reductionists would be able to handle the problem easily by demanding an explanation for the steps that take us from one of these dichotomous areas to the other. If these areas are absolutely and ontologically separate, one cannot scientifically explain the "passage" from the nonliving to the living; that is, one cannot explain the advent of life. Furthermore, one cannot scientifically explain how one passes from living organisms to conscious organisms and to self-conscious human beings.

We agree with a lot of what Gerd Sommerhoff and others have said: "Even if we knew what happens within the smallest cellular structure present in a living organism, we would still find ourselves faced with the fact that a living system, as an organized entity, expresses, due to the distinctive nature of its organization, unique forms of behavior that must be studied and understood at their level, and that is because the importance of all living things depends on this."[19] In particular, we agree that "the organized entity expresses, because of the distinctive nature of its organization, unique forms of behaviour which must be

---

[19] Sommerhoff, "Le caratteristiche astratte dei sistemi viventi," 164.

studied and understood at their level,"[20] although we do not completely concur with the phrase "because the importance of all living things depends on it."[21] We also completely disagree with exclusively attributing such characteristics only to living organisms, since we wish to generalize this systemic approach and extend it to nonliving organized structures, such as those studied by chemists. The view expressed by Sommerhoff has been noncontroversially accepted but only with regard to two types of systems: the living organism and human consciousness. However, we argue that this principle is valid for all organized systems. The main examples we will offer to defend this position are molecular and macromolecular systems.

It is within this perspective that we will frame the journey that has led from the discovery of the atom to the plurality of atoms, molecules, and macromolecules, all the way to the threshold of life. Rather than embracing the separation between the nonliving, the living, and the human, which has led to dichotomies that undermine scientific knowledge and philosophical understanding, our systemic approach argues that systems do not conform to such dichotomies. We will demonstrate that many systemic properties that are attributed, for example, to the conscious brain are also present in molecules and macromolecules and, thereby, transform these dichotomies into epistemological differences to be studied and analyzed.

Finally, we wish to highlight one aspect of scientific explanation that stems from the systemic approach, that is, the idea that explanation through laws is no longer the only possible type of scientific explanation. This is evident in biology, in which many scientists have wondered whether one can speak of specific laws, but it is also evident in chemistry. As Villani has argued extensively elsewhere,[22][23] chemistry does not seek to discover chemical laws or to develop explanations in terms of chemical laws but, rather, to explain chemical entities, their properties and their interactions, and their transformations.

[20] Ibid.
[21] Ibid.
[22] Villani, *La chiave del mondo. Dalla filosofia alla scienza: l'onnipotenza delle molecole.*
[23] Villani, *Complesso e organizzato. Sistemi strutturati in fisica, chimica, biologia ed oltre.*

1

Qualitative Atomism and Life within the 18th-Century Atomistic Perspective

Since the time of the ancient Greek cosmologists, the problem regarding the investigation on the nature of matter has been approached from either a local and reductionist perspective or a global and holistic perspective. When examining these two contrasting approaches, it is useful to do so from the point of view of mereology, which is the branch of philosophy concerned with understanding and defining the nature of the relation between wholes and their parts. From a mereological standpoint, the local and reductionist approach considers that the nature and properties of a whole are entirely defined and determined by the nature and properties of its distinctive parts. When distinguishing different levels of microscopic or macroscopic scale, the reductionist will claim that the properties at the higher ontological levels are entirely deducible from and reducible to the properties at the lowest ontological level. In this view, causation is thought to occur only in an upward direction, from the lower-level processes and reactions to the higher-level phenomena. Therefore, the reductionist completely rejects the possibility of downward causation, that is, the idea that the whole or the environment in which the parts exist and are related has an effect on those parts and their relations. Thus, according to reductionism, the whole is nothing more than the summation of its constituent parts.

The global and holistic perspective takes the exact opposite approach to reductionism. According to the holistic perspective, the world as it is consists only of wholes existing in various relations to one another and embedded within an even more complex environment. According to this view, it is when we attempt to understand what governs the properties and behaviors of wholes that we conceptually divide them into parts, some of which are more fundamental and some less fundamental, depending on the level of properties and reactions on which a given science focuses. However, cutting the world at the seams is an ad hoc conceptual exercise governed by whichever model of fundamental nature dominates a particular science at a given time. Although the holistic perspective does not deny the ontological existence of parts and lower-level processes, it considers that the causal effect and significance of the parts are completely absorbed by and disappear within the global context of the whole. Thus, from this point of view, not only is the whole greater than the mere summation of its parts, but it

From the Atom to Living Systems. Marina Paola Banchetti-Robino and Giovanni Villani, Oxford University Press.
© Oxford University Press 2023. DOI: 10.1093/oso/9780197598900.003.0002

is also more ontologically and causally significant than the parts that, once the whole has emerged, cease to have any ontological and causal distinctiveness of their own. Thus, when not properly checked, the global and holistic perspective runs the risk of, at least tacitly, negating the ontological dependence of the whole upon its constituent parts and of privileging downward causation at the cost of downplaying upward causation.

The systemic nonreductionist perspective that we are proposing in this book, however, takes a balanced and science-based approach that considers the insights of both the local and global perspectives, while addressing the weaknesses of taking either approach exclusively at the expense of the other. The perspective that we propose approaches the nature of matter by studying the relations be-tween the parts, the structure according to which the parts are arranged, and the dynamic relations between the parts and whole of which they are constituents. From a dynamic perspective, one must also consider the relations between the objects of study and the environment in which they are located and within which they form part of an even greater whole. Albeit ontologically dependent on the properties of their constituent parts, the properties of wholes are not thought to be deducible from or entirely reducible to the properties of those parts. Both the reductionist and holistic approaches have held favor with scientists and philosophers at different moments of our intellectual history and have often been intimately connected with philosophers' and scientists' overall conception of the cosmos, of nature, and of its inner workings.

Although our investigation in this work begins with the 17th century, during which time modern science as we know it was first being conceptualized and developed, it is important to examine the ancient origins of the early modern conception of matter upon which we will be focusing in Chapter 2. More pre-cisely, the early modern atomism of philosophers such as Pierre Gassendi, and even the corpuscularianism of René Descartes and Robert Boyle, resulted from the Renaissance revival of ancient atomism in Europe that occurred with the rediscovery of Lucretius's *De Rerum Natura*, in which the Roman philosopher extolls the virtues of the particulate conception of matter that he inherited from Epicurus, Leucippus, and Democritus.

As we examine ancient and modern atomism more closely, however, we note the beginning of a divergence between these two positions. For example, an-cient atomists drew no distinction between the atoms of animate and of inan-imate matter. On the other hand, many early modern thinkers affirmed this dichotomy. For example, one of the cardinal principles of Cartesianism was precisely the dichotomy between animate substance (*res cogitans*), that is, mind or consciousness, which is nonmaterial and not extended in space and inert or inanimate material substance (*res extensa*), whose only property is extension, from which all other properties derive. Physical bodies are simply inert matter according to Descartes and, although he recognized the existence of complex machines or "automata" that could simulate the behavior of conscious and living beings precisely because of their complexity, he did not consider them to be animate or alive in any proper sense due to their lack of consciousness. In fact, Descartes did not consider plants or even nonhuman animals to be properly living things because he believed that they lacked consciousness. For him, nonhuman animals were simple "automata," since he thought that all of their behaviors could be fully explained simply by appealing to the laws of mechanics without any reference to consciousness. He considered that only human beings were living things because much of human behavior, including thought, introspection, and self-awareness, could be explained only by positing the existence and presence of consciousness. In response to this Cartesian view, the 18th-century materialist philosopher Pierre Louis Moreau de Maupertuis precisely opposed this dichotomy between animate and inanimate matter by assigning to all atoms some principle of intelligence, memory, and the like, because he and other materialist philosophers understood the impossibility of explaining the manifestation of life and consciousness by postulating the existence of only strictly inert particles of matter.

The work of Maupertuis stands at the dividing line between natural history and physiology and between structuralism and functionalism. Although his ideas could never correctly be termed "vitalistic," they do share some features with medieval and Renaissance vitalism, and they constitute an attempt at conceptualizing material particles as something more than simple inert matter. However, his sophisticated perspective regarding the distinction between inert and living or animate matter was not the first attempt to endow strictly corporeal bodies with more than simply mechanistic qualities.

In the context of medieval, Renaissance, and even early modern alchemy, particulate theories had coexisted quite comfortably with vitalistic conceptions of matter. In this regard, the particles posited by Maupertuis resembled more the *semina rerum* of medieval and Renaissance vitalistic conceptions of matter, material seeds that were endowed with an internal life force, than they resembled the inert particles otherwise implied by strict materialism. We shall now turn to some of these vitalistic views in order to locate the antecedents of the antimechanistic, albeit materialist, perspective exemplified by Maupertuis and other materialists such as Diderot.

The first of these views that we shall examine is the hylomorphic atomism of the early modern alchemist Daniel Sennert, since his theories constitute the first important attempt to integrate the vitalistic perspective with the atomism that dominated early modern natural philosophy. We shall see that Sennert made a number of important theoretical and methodological contributions to the development of late-16th and early-17th century chemistry, including developing an early version of chemical structuralism and using experimental means to confirm chemical theories.

Qualitative Corpuscularianism in the Renaissance and Early Modern Periods

Historians of science once held that all atomistic theories of matter, whether in their ancient or their modern iterations, were inevitably associated with a mechanistic account of the nature of particles and of the relations between them.[1] However, recent historiography has established that there was a continuity of atomistic thought from the 15th to the 19th centuries[2] and that some of these theories were framed within a vitalistic cosmology and qualitative conception of particles. The 17th-century mechanistic atomism of Gassendi was made possible not only by the rediscovery of Democritean atomism but also by the survival of the particulate theory of matter through late antiquity and into the Middles Ages and Renaissance.

Although ancient atomism had been strictly materialistic, mechanistic, and deterministic, the many advocates of corpuscularianism during the Renaissance attempted to reconcile it with a nonmechanistic conception of the universe. They accomplished this by hypothesizing that fundamental particles of matter contained within them an immaterial formative principle and thus developed a vitalistic and qualitative type of atomism. One of the most important representatives of this view was the Renaissance philosopher Giordano Bruno, although, for the specific purpose of our discussion, the more relevant figures are the early modern alchemists Daniel Sennert and Jan Baptista van Helmont. Their ideas and work anticipated much of early modern chemistry and exerted an invaluable influence on important chemists such as Robert Boyle and his contemporaries. They will be discussed later in this chapter.

For the moment, it is important to emphasize that the attempt to identify the fundamental constituents of nature had a profound, though not always successful, impact on the development of early modern chemical philosophy. Additionally, besides considering the role of philosophy and metaphysics in the development of early modern chemical philosophy, one must also consider the role played by the concrete practices of the laboratory. It is the importance of experiment that ultimately led to the rejection of Aristotelian hylomorphism and the accompanying Scholastic notion of substantial form. Both concepts had

[1] Stone, "The Atomic View of Matter in the XVth, XVIth, and XVIIth Centuries," 445.
[2] Ibid.

a stronghold on the study of natural philosophy throughout the Middle Ages, Renaissance, and well into the 16th century.

We will, therefore, briefly discuss both hylomorphism and the theory of substantial form and show that, whatever function these concepts served within chemical explanations, that function would later be better served by the structuralist approach to understanding chemical wholes. As we shall see later in this chapter, long before the development of chemical structuralism in the 19th century, Sennert suggested that the limitations of the Scholastic theory of substantial form for explaining chemical processes could best be addressed by introducing the concept of structure within the hylomorphic perspective.

The Rejection of Hylomorphism and the Theory of Substantial Form

The Scholastic theory of substantial form was closely associated with the concept of substance as it had been articulated within Aristotelian metaphysics. The concept of form, for Aristotle, is a fundamentally mereological concept; that is, it must be understood as a theory of parts and wholes precisely because Aristotle was trying to resolve the question of what it is that makes the whole something more than simply the additive summation of its parts. One of the reasons for asking this question relates to the fact that wholes have a unity and distinctive identity, through time and through changes, that is "something over and above" the simple collection of their parts. Here, Aristotle is concerned with what it is that makes a whole, whether it be a simple whole or a composite whole, and how it can have an internal unity that is distinct from mere heaps of objects that have no internal unity. For example, what makes a piece of marble a unified whole as opposed to a pile of rocks, which is not a unified whole? Put more simply, what is the "something over and above" the parts that is present in the unified whole but is not present in a heap of objects? The answer, according to Aristotle, is that the block of marble has substantial unity, while the heap does not, and that the unity of the block of marble exists by virtue of its substantial form.

This theory, which is called hylomorphism, conceives of unified and stable substances as the combination of matter and form. For Aristotle, forms serve as the organizing principles that account for the unity, identity, order, and essential properties of things, whether these things be simple wholes or composite wholes. However, what exactly is substantial form for Aristotle? Since he never actually answers this question, beyond saying that it is a unifying principle, 12th-century Scholastic philosophers attempted to develop a more satisfactory conception of form. For them, substantial forms are immaterial principles that are responsible for endowing otherwise undifferentiated matter with distinctive properties. In the context of medieval chemical practice, this theory had a profound impact on the Scholastic theory of mixts, as natural philosophers attempted to explain why a homogeneous mixt would display different properties from those of its components.

According to Thomas Aquinas, for example, when two substances are mixed, each of those substances loses its form and its distinctive properties and a new form is produced that accounts for the properties of the homogeneous mixt. Aquinas's view implies that the component substances of a genuine mixt can never be recovered through chemical analysis since their substantial forms no longer exist. Thus, a true synthesis (*synkrisis*) precludes the possibility of analysis (*diakrisis*). Aquinas would explain any kind of analysis by saying that, if the attempt to analyze a mixt is successful at recovering the individual ingredients, then the mixt clearly did not have a substantial form and was thus not a genuine and homogeneous mixt.

The idea that genuine mixts could not be chemically analyzed into their constituents, however, contradicted experimental observations. Thus, early modern chemists ultimately rejected the Scholastic theory of substantial form and sought to find a suitable alternative that would explain the identity, unity, and stability of substances while also allowing for the possibility of successfully analyzing genuine mixts. In the discussion that follows, we shall see that each chemist's proposal for an alternative to the theory of substantial form would ultimately define the core of that chemist's ontology and chemical philosophy.

The development of an epistemology of experiment by early modern chemists significantly impacted on the abandonment of hylomorphism, whether in its Aristotelian or its Scholastic iteration. This, however, was a gradual process and early attempts did not so much seek to eliminate substantial form as to revise this idea to coincide more with experimental observations. An important figure in this transition is the German experimental alchemist Daniel Sennert (1572–1637). Sennert's work is also particularly interesting because it represents an example of a chemical philosophy that is both vitalistic and atomistic, in addition also to representing an early attempt at conceptualizing chemical wholes in structural terms, within a generally hylomorphic framework.

Sennert rejects Aquinas's conclusion that true synthesis must preclude the possibility of analysis because this notion is clearly inconsistent with experimental evidence. However, his desire to revise hylomorphism also stems from his attempt to articulate a distinction similar to that made, in the following century, by Denis Diderot and the French materialists. This is the distinction between *la matière* and *les matières*, that is, between theoretical undifferentiated material substance and the diverse concrete substances of the chemist's experimental laboratory.[3]

---

[3] Pépin, *La philosophie expérimentale de Diderot et la chimie.*

Like the French materialists after him, Sennert wishes to articulate a notion of substance that better captures the concrete experiences and practices of experimental chemistry rather than the abstract speculative conception favored by the Scholastics. In this regard, Sennert also anticipates the later approach taken by Lavoisier, who rejects purely abstract speculations about the nature of matter in favor of an empirical and quantitative approach for identifying elementary substances. Furthermore, this distinction between how one conceives of fundamental matter and how one identifies diverse and varied chemical substances anticipates the distinction made centuries later by Friedrich Paneth between the a priori and abstract concept of basic substance and the empirical notion of simple substance.[4]

Sennert is a vitalist who embraces hylomorphism, but he also believes that atomism provides a better theory of mixture, although he disagrees with the mechanistic aspects of ancient atomism. Because of this, he develops a hybrid theory that reconciles the particulate theory of matter with vitalism and hylomorphism. For Sennert, the different types of atoms that compose a substance are endowed with distinctive immaterial forms that define their essence and their distinctive properties. Thus, it is the particles themselves, rather than the substances, that have form. Additionally, for Sennert, atoms do not simply have the mechanistic properties of mass, shape, spatial location, and motion or rest but also have distinctive essential properties that determine them as atoms of a specific type. The essential properties of an atom are themselves determined by the distinctive form of that atom. Thus, using sodium and chlorine as examples, Sennert would say that an atom of sodium has a distinctive substantial form and, as a result, it has distinctive essential properties that make it an atom of sodium and different from an atom of chlorine, which also has its own distinctive form and distinctive properties.[5]

Sennert is unique among early modern chemical practitioners in theorizing that the atoms of a particular substance do in fact have distinctive properties that account for the unique properties of that substance as a whole. It is interesting to note that, although his chemical ontology and general theory of hylomorphic atomism contrasts radically with the chemical atomism of John Dalton, Sennert seems to be anticipating the turn that Dalton would make in the 19th century toward arguing that atoms of different elements do have different properties and account for the unique properties of those elements. In this regard, Sennert's chemical philosophy was centuries ahead of its time.

---

[4] See Paneth, "The Epistemological Status of the Chemical Concept of Element," 1–14 and 144–160.

[5] Sennert, *De chymicorum cum Aristotelicis et Galenicis consensu ac dissensu*. See also the discussion of Sennert's experiments in Newman, *Atoms and Alchemy: Chymistry & the Experimental Origins of the Scientific Revolution*.

Once Sennert proposes that atoms of distinctive substances have unique forms and properties, the question that remains to be answered is what happens when the atoms of different substances, such as the atoms of sodium and of chlorine, come together to form a mixt such as sodium chloride (NaCl). In other words, Sennert still needs to explain why the resulting mixt has such different properties from those of its constituent atoms. This is another point at which Sennert's hylomorphic atomism acquires a distinctively modern tone, for, in order to explain how mixts acquire their distinctive properties, he develops a very early version of chemical structuralism.

According to Sennert, although simple atoms are fundamental particles each of which is endowed with its own distinctive substantial form that determines the type of atom that it is, when atoms of different types come together in a mixt, they arrange themselves according to distinctive structures to form unified and stable chemical wholes having different properties from those of their more fundamental parts. Thus, the observable properties of the mixt, which differ from the properties of its constituents, are not the result of individual atoms losing their substantial forms. Rather, it is the structural arrangement of these atoms that creates the form of the mixt and accounts for its distinctive properties that may differ from those of its constituent parts. Thus, although both sodium and chlorine are poisonous gases, sodium chloride is a nontoxic substance with a crystalline structure.

According to Sennert, the substantial forms of the atoms of sodium and of chlorine have not changed or been destroyed within the mixt. Rather, it is the structural arrangement of such atoms that results in the unique properties of the mixt. Sennert, however, does not equate structural form with substantial form, since the former is an accidental property resulting from bringing together distinctive atoms, while the latter is an essential property of the atoms themselves. Thus, the mixt does not have its own substantial form, while its constituents do, and this explains why the constituent atoms are recoverable with their original properties through the process of chemical analysis. Additionally, Sennert considers that the chemical wholes within which atoms are arranged are nonfundamental chemical species or, to use Gassendi's term, *molecules*, whose specific properties are empirically observable. Therefore, his theory involves an interesting interplay between essence, as represented by atomic substantial form, and structure because it is the structural arrangement of atoms coupled with the distinctive substantial forms of those atoms that account for the specific observable properties of substances.

As an experimental chemist with little sympathy for pure speculative metaphysics, however, Sennert concedes the difficulty of positing any sort of substantial form to explain the essential properties of material substances, whether the substantial form is posited as belonging to the substance as a whole or to each of its atomic particles. Yet, he also believes that experimental evidence can be garnered to support such a hypothesis, once it can be demonstrated that the constituent ingredients of a chemical mixt are recoverable through analysis. For him, the possibility of successful analysis confirms the idea that the constituent atoms have retained their identity. However, if it can also be shown that the constituent substances cannot be further analyzed but retain their stability and identity, then this would confirm the notion that the atoms have the identity that they have due to indestructible substantial forms. Note here that Sennert's view is still limited because he has no working paradigm to account for chemical identity and stability other than the concept of substantial form. Although this is certainly a weakness of his account, one can at least appreciate his attempt to move beyond the limitations of the Scholastic theory.

To provide experimental support for his structural hylomorphic atomism, Sennert developed several variations of the "reduction to the pristine state," an experiment that Robert Boyle would later appropriate but for an opposite purpose, that is, to confirm his own theory of mechanistic structure against any and all conceptions of substantial form. The goal of the reduction to the pristine state is to establish that constituent substances are indeed recoverable from homogeneous mixts. Although pre-Sennertian alchemists and iatrochemists had availed themselves of this experiment, they had used it for the strictly pragmatic objective of purifying metals. It is historically and scientifically significant that Sennert is the first chemical practitioner to use this analytical procedure to provide empirical confirmation for a theoretical hypothesis. Sennert justifies using this experiment because, for him, "the validity of chymical analysis is borne out of the Scholastic 'axiom' [that] 'The things into which composites can be dissolved are the things of which they are made'—Based on Aristotle's *De caelo* 3 302a 15–18."[6]

Even though Sennert performs the reduction to the pristine state using various mixtures, the most well known of his reductions involved dissolving silver in *aqua fortis* (nitric acid), which would then result in a clear blue liquid. He then filtered the resulting liquid to establish that no solid residue remained and that the mixt was indeed homogeneous. Next, Sennert precipitated the silver back from the liquid by pouring alkaline salt of tartar (potassium carbonate) into the solution. After cleaning, drying, and heating the precipitate, he observed that a good quantity of the silver had been recovered from the solution, thereby falsifying Aquinas's claim that the components of a homogeneous mixt are not recoverable via chemical analysis. In falsifying Aquinas's view on analysis, Sennert also believes he had falsified the Scholastic theory of substantial

---

[6] Newman, *Atoms and Alchemy: Chymistry & the Experimental Origins of the Scientific Revolution*, 97.

form, as applied to whole substances rather than to their component particles. He explains that

> the metal preserves its own unchanged nature in [the mixt] and is easily precipitated at the bottom in the form of a very fine powder, which may then be reduced again into the metal. Thus also if a single mass be made by fusion of gold and silver together, and they come together through the smallest atoms (*per minima atomos*) to the degree that no one may recognize that this body consists of different components, still each metal retains its own form in those minimal atoms, and can be separated by aqua fortis and can be reduced into its original body.[7]

For Sennert then, the atomistic implications of this experiment are clear. Although the silver was completely dissolved in the acid, it could easily be recovered to a great degree of quantitative precision. Believing that the atoms of the silver had retained their substantial form in the liquid solution, thereby explaining why the original substance could be recovered, Sennert would explain the various changes throughout the experiment as follows. Although the atoms of the silver retained their substantial forms in the liquid solution, they do not manifest the properties of the original silver because the structure that held them together in the silver has been dissolved by the *aqua fortis*. Once the salt of tartar is added to the solution, however, the original structure of the atoms is reconstituted, thereby manifesting the silver once more with all its properties intact. For Sennert, the results of this experiment confirm the theory that the atoms of the silver have retained their identity in the liquid solution and, if this is the case, they must have retained their substantial forms as internal principles of unity.

Sennert's theory is, in many regards, much more sophisticated than the Scholastic view and, as already suggested, is also prescient of later developments in modern chemistry. As Bensaude-Vincent and Stengers have explained, although the reduction to the pristine state cannot ultimately prove Sennert's compositional theory of matter, "the operations that brought reversibility to the fore [in chemistry] inspired a new classification of chemical procedures that theoretically favored purification processes."[8] In addition, Sennert's hylomorphic atomism serves as an important transitional hypothesis from the Scholastic theory of substantial form to the mechanistic corpuscularianism that will be discussed in the next chapter. Finally, his compositional theory

---

[7] Sennert, *De chymicorum*, as translated and cited in Newman, *Atoms and Alchemy: Chymistry and the Experimental Origins of the Scientific Revolution*, 362.
[8] Bensaude-Vincent and Stengers, *A History of Chemistry*, 32–33.

shows "the interplay between the structural and the substantial types of explanation. He uses a generalized microstructural explanation in combination with
chymical properties originating in the substantial forms to explain the origin of
macrolevel effect[s]."[9] His ideas are, therefore, innovative and anticipate the work
of later chemists such as Boyle but also the eventual structural focus of modern
chemistry.

It is also important to note that Sennert is one of the first chemists to attempt
a quantitative determination of atomic dimensions. In 1636, he devised "a series
of experiments designed to prove the existence of atoms [and] described a distillation in which a stream of alcohol vapor passed through a sheet of paper,
the density of which was supposed to give an idea of how small the atoms really were."[10] Thus, in this regard, Sennert is clearly anticipating the quantitative turn that modern chemistry would ultimately take in the 18th century. As
suggested earlier, Sennert's work also presciently involves the interplay between
theory and experiment "in that, in the reduction to the pristine state, he uses
a paired cycle of synthesis and analysis within the context of a demonstrative
proof. Unlike alchemists and iatrochemists before him, Sennert is using specific
experiments not for simply practical ends but for theoretical purposes to corrob-
orate his hylomorphic atomism. His approach would have striking repercussions
in the early modern period,"[11] such as when Boyle uses experiments like the redintegration of potassium nitrate to provide empirical support for his mechanistic
corpuscularianism.

Of course, in spite of his theoretical sophistication, we must also recognize
that Sennert was a man of his time and that he was committed to the hylomorphic paradigm that he tried to reform but within which he was still immersed.
Despite the prescience of his structural atomism and his analytical methods, his
conception of fundamental matter is still firmly grounded in the Aristotelian and
Scholastic notions of substance that would later be rejected by the mechanistic
philosophy. Additionally, his hylomorphic atomism retains significant elements
of the vitalism that dominated the traditions of medieval and Renaissance alchemy. Finally, although he turns to experiment to support his theory of matter,
Sennert fails to sustain the notion of corpuscular substantial form within chemical explanations. As Bensaude-Vincent and Stengers have stressed, this failure
occurs because his reduction to the pristine state is "not sufficient to support
his atomism, since the Scholastic idea of the mixt would be perfectly capable of
explaining reversibility."[12] For instance, a Scholastic alchemist could argue that

---

[9] Newman, *Atoms and Alchemy: Chymistry and the Experimental Origins of the Scientific Revolution*, 136.
[10] Meinel, "Empirical Support for the Corpuscular Theory in the Seventeenth Century," 78.
[11] Newman, "What Have We Learned from the Recent Historiography of Alchemy?", 6–7.
[12] Bensaude and Stengers, *A History of Chemistry*, 32–33.

the clear blue solution that results from pouring *aqua fortis* into the silver is not a genuine mixt, precisely because the silver is easily recoverable by adding salt of tartar. Thus, Sennert's experiment does not confirm his theory. Philosophers of science would point out that this example illustrates the underdetermination of theory by evidence, since the recovery of the silver from the liquid solution can be explained by competing and incompatible theories.

We now turn to the qualitative atomism of Jan Baptista van Helmont, since this constituted the second major attempt to conceptualize atomism within a vital-istic conception of nature. Helmont is important for many reasons, not the least of which are his contributions to the ontological theory of disease. Although we shall address fermentation in much greater detail in Chapter 9, it is important to note here that, once atomism and corpuscularianism became fully mechanistic, it became impossible to account for phenomena such as fermentation because, in the 16th and 17th centuries, fermentation was still considered a vitalistic phe-nomenon. Thus, Helmont's attempt to conceptualize fermentation as a necessary process in the production of both living and nonliving bodies from seeds is a bold attempt to render early modern atomism heuristically capable of explaining the possibility of life.

Jan Baptista van Helmont and Qualitative Atomism

In order to examine qualitative atomism in Helmont's work, it is imperative to clarify the importance of vitalism played in 15th- and 16th-century alchemy. Vitalism is the view that "vital forces" or "vital spirits" have a causally operative role in nature. Vitalistic descriptions of natural phenomena tend to be qual-itative, and vitalistic processes are regarded as being holistic and purposive. However, vitalism was not incompatible with a particulate conception of matter. In fact, Renaissance and early modern alchemists often embraced conceptions of matter that were both corpuscularian and vitalistic, and the alchemist Jan Baptista van Helmont (1580–1644) was no exception. One of the central issues informing 15th- and 16th-century alchemy, as well as early modern chymistry, was that of how to interpret the concept of vital spirits. Certainly, many vitalists interpreted this idea in immaterial terms, very much in the same way that many Scholastic philosophers interpreted the notion of substantial form immateri-ally. However, by the end of the 16th century, the concept of vital spirits became unambiguously chemical, although it strongly retained the vitalistic elements it had acquired in the work of figures like Paracelsus, Girolamo Frascastoro, Isaac Beeckman, and Sebastien Basso. Although Joseph Duschene and Oswalt Croll reinforced this chemical interpretation when they stated that "the only active remedies [are] those prepared by using spirits extracted by distillation,"[13] such an unambiguously chemical interpretation of vital spirits is even more evident in the work of Helmont.

What is even more interesting is that Helmont's chemical conception of vital spirits is solidly grounded in an, albeit qualitative and vitalistic, atomism. Although mechanistic corpuscularianism and atomism had already taken hold during his lifetime, Helmont rejects the explanatory value of strictly mechanistic explanations. For example, he claims that mechanistic explanations for chemical processes such as the mixture of substances would be restricted to considering only the mechanistic properties of shape, size, and motion of fundamental particles. Under such a model, a mixture of substances would have to be explained as the mere juxtaposition of physical parts. It is interesting that Helmont's misgivings here describe precisely the concerns that would lead to the later development of structural chemistry, an early version of which was central to the work of Robert Boyle, in order to account for analysis and synthesis precisely within a mechanistic framework. According to Helmont, however, a "purely mechanical juxtaposition of [physical] parts does not bring about a real mixture [of substances]"[14] because a physical juxtaposition does not yield a true synthesis. Therefore, strictly mechanistic principles cannot adequately explain or predict chemical reactions.

Against mechanistic explanations, Helmont postulates that chemical reactions, mixtures of substances, and transmutations depend on ferments that are contained in *semina*, formative principles from which all natural bodies were said to originate. These ferments were themselves considered as formative spiritual agents and, for Helmont, although chemical analysis and the "reduction of bodies into their *minima partes* is a 'pre-condition' for transmutation—[this] is ultimately a spiritual process."[15] Helmont explains many physical changes in this qualitative manner by appealing both to atomism and to the notion of *semina* as vitalistic formative principles. For example, he interprets the production of gases as involving "the disposition of the *tria prima* [salt, sulfur, and mercury] within the corpuscles of water. . . . The purely material change, that is the attenuation of water parts into atoms, is preliminary to a process that is qualitative, not mechanical."[16]

Another example of this approach is his corpuscular explanation of certain chemical reactions, such as the transmutation of iron into copper and the production of glass, since "both [chemical reactions] are explained in terms of

---

[13] Clericuzio, "The Internal Laboratory," 53.
[14] Clericuzio, *Elements, Principles and Corpuscles: A Study of Atomism and Chemistry in the Seventeenth Century*, 58–59.
[15] Ibid., p. 60.
[16] Ibid., 57–58.

addition and subtraction of particles. The notion of atoms is also employed in van Helmont's theory of mixture and generation. . . [although he] imposed severe restrictions on the corpuscular theory of matter. [This is because] *semina rerum* and ferments are the active principles on which all natural phenomena ultimately depend."[17]

Returning now to the concept of vital spirit, Helmont conceives of it in chemical terms as an alkaline volatile salt that moves through the body and that is "generated from the volatile salt contained in cruor [blood without spirit] and by means of a local ferment."[18] This idea of ferment links Helmont's theory of vital spirit to his corpuscularian theory of matter, since it intersects atomistic explanations of chemical phenomena with certain aspects of Paracelsian vitalism.[19] For Helmont, atoms as physical units undoubtedly have qualitative determinations, not simply mechanical properties.[20] He considers that *semina*, as formative principles, work together with atomic particles to bring about changes in nature by providing the spiritual and vital force of action that brings about qualitative chemical alterations. That being said, Helmont moves beyond Paracelsus and other predecessors to the extent that he also employs quantification as a tool for research and brings to bear observational evidence gathered from his own experiments to support the central claims of his chemical philosophy.[21] Therefore, in spite of his lingering commitment to a vitalistic conception of matter, Helmont's use of quantification and his efforts to support his speculative claims regarding *semina* and vital spirit with experimental evidence place his work squarely in the modern tradition of Galilean science. However, Helmont's importance in the context of early modern science goes further, inasmuch as he takes one of the first steps toward the naturalization of chemical explanations by postulating physical particles as the objects of chemical change. Given that the framework of Helmont's explanations was not entirely physicalistic and that he embraced a theologically centered cosmology, Helmont's contribution to the naturalization of chemical philosophy was probably inadvertent. Nevertheless, the significance of his ideas for the modernization of chemical theory and practice should not be underestimated.[22]

Before we move to a discussion of the 18th-century mechanistic theory of matter in the next chapter, it is important to end our discussion of vitalistic

---

[17] Ibid., 58–61.

[18] Clericuzio, "The Internal Laboratory," 53.

[19] Banchetti-Robino, "Ontological Tensions in 16th and 17th Century Chemistry: Between Mechanism and Vitalism," 173–186.

[20] Clericuzio, *Elements, Principles and Corpuscles: A Study of Atomism and Chemistry in the Seventeenth Century*, 56.

[21] Debus, *The Chemical Philosophy*, 327.

[22] Banchetti-Robino, "Van Helmont's Hybrid Ontology and Its Influence on the Chemical Interpretation of Spirit and Ferment."

atomism with the attempt, by the French materialist Maupertuis, to formulate an explanation for the possibility of life and consciousness within the framework of materialistic atomism. What we shall see, in this discussion, is that the challenges faced by materialism in explaining how life and consciousness are possible at all ultimately led Maupertuis to embrace a type of vitalism, while retaining the façade of a materialist ontology.

The French Materialists and the Concept of Living Points

As mentioned at the beginning of this chapter, once the vitalistic atomism of natural philosophers such as Helmont and Sennert was completely supplanted by the mechanistic theory of matter, it became that much more difficult to explain both how life and consciousness are possible and how inanimate matter such as food can become part of animate bodies. For mechanistic philosophers such as Descartes, life and consciousness in human beings were explained by taking a dualistic position on the subject of substance and by postulating that human beings are "ghosts in the machine." That is, human beings are endowed with a rational mind that is located in the pineal gland of the brain, which is itself part of our physical and mechanistic body. For Descartes, the rest of the physical universe is strictly made of inert matter, so that material substance is devoid of any active forces or intrinsic principles of motion. As mentioned earlier, Descartes denies that nonhuman animals are true animals; that is, they lack "anima" or consciousness, and so they are to be regarded as purely mechanistic bodies. All the properties, interactions, or motions we observe between nonhuman material bodies are explained strictly by the collisions of fundamental corpuscles whose only properties are shape, size, and spatial location and whose motions and collisions are governed by deterministic laws of mechanics.

Although the authors of this book do not agree with Descartes' assessment that animals are mere automatons, his position can at least in part be understood due to his application of Ockam's Razor, that is, the principle requiring that one should always opt for the simplest explanation, that is, the explanation that requires the least amount of postulates without losing heuristic value. Applying this principle, Descartes believes that the movements and "behaviors" of nonhuman animals can be explained entirely by appealing to the collisions of particles and the laws of mechanics, without needing to appeal either to the presence of consciousness or even of life. On the other hand, Descartes believes that such purely mechanistic explanations do not suffice to explain human behavior because, from a subjective and introspective perspective, one's consciousness and thought processes are absolutely phenomenologically primary and undeniable. However, Descartes' arguments for substance dualism suffer from a pernicious circularity and remain unconvincing, particularly because they fail to explain how an extended and inert substance (matter) can ever interact with a nonextended, active, and thinking substance (mind) without violating the very laws of mechanics that deterministically govern all material bodies.

Several philosophers reacted to the failures of Cartesian dualism by adopting substance monism, that is, by postulating the existence of only one substance. While the alternatives of Berkeleyan idealism and Spinozist monism were appealing to few philosophers, materialist monism was embraced by a sizeable number of thinkers, particularly by a group of French philosophers clustered around the intellectual circle of the Baron d'Holbach, who was one of the leading materialist philosophers of the 18th century. Among these thinkers were Julien Offray de la Mettrie, Claude Adrien Helvétius, Denis Diderot, and the Newtonian and anti-Cartesian philosopher, Pierre Louis Moreau de Maupertuis. Some of these thinkers, such as La Mettrie, were willing to embrace the full implications of materialist monism and, as he proposes in his important 1747 work *L'Homme Machine* (*Man a Machine*), even human beings are simply unthinking automatons governed by deterministic laws of motion. However, Maupertuis remains unsatisfied with La Mettrie's proposal, especially since he is concerned with explaining both how life is possible and how food becomes part of animate bodies. Maupertuis is also very interested in explaining how the characteristics of parents are passed on to their offspring, not only in the case of human beings but also of nonhuman animals. The desire to explain such phenomena ultimately leads him away from Newtonian atomism, with its strongly empirical and mathematical grounding, and toward the more speculative metaphysics of Leibniz. In order to explain phenomena such as consciousness, life, and epigenesis, Maupertuis postulates what can be called organic monads or "living points," and he attributes these to all fundamental constituents of matter.

Maupertuis proposes an ontology of living points to explain both the generation and the reproduction of matter, and he conceptualizes this idea on the Leibnizian concept of monad and model of perception, although we must keep in mind that Maupertuis never diverges from his fundamentally materialist ontology. As Ernst Cassirer stresses, "Maupertuis does not attempt to conceive the fundamental units of which all natural processes exist as metaphysical points— as Leibniz had done—but as physical points. In order to arrive at these units we need not depart from the corporeal world as such, but we must enlarge the concept of matter in such a way that it does not exclude the basic facts of consciousness."[23] At the beginning of his *Système de la nature* (1751), Maupertuis affirms what he takes to be the heuristic value of his conjectures regarding living points and states that, with his system, "all of the unsurmountable difficulties within

---

[23] Cassirer, *The Philosophy of the Enlightenment*, 88.

other systems will disappear,"[24] such as the question of how living things can exist if they are made only of simple atomistic particles or how children inherit their parents' traits. For example, he postulates that epigenesis is due to a kind of memory that the matter retains of its previous condition, which makes it possible for this previous condition to be passed from the parent to the offspring. "[T]he elements which convene to form the fetus," he affirms, "swim in the seeds of the paternal and maternal animals. Each one, however, extracted from a part similar to the one it must now form, preserves a kind of memory of its former situations, and will resume it as often as it can, so as to form an equal part in the fetus."[25]

Maupertuis conceives of the elements that form the fetus as being homogeneous and nonorganic seminal matter, and he refers to Parisanus's *De subtilitate* (1623) as the source of the idea that, within each fundamental particle or element of living beings, there is a disposition to manifest a form that is already contained within itself. This idea, of course, also harks back to the medieval and Renaissance concept of seminal reasons (*logoi spermatikoi* or *rationes seminales*), which were thought to be formative principles within the seeds of things. According to Maupertuis, "generation occurs through a process in which a formal and internal seminal principle of organization progressively modifies a material substratum, producing differentiations in succession, organ after organ, ultimately producing a body whose final structure mirrors its own."[26]

Maupertuis elaborates this process of aggregation by appealing to the Newtonian-based chemistry of Étienne François Geoffroy, availing himself of Geoffroy's distinction between the substantial and the relational character of chemical affinities. Maupertuis is not satisfied with merely saying that a fetus is formed when two seeds are mixed, since this lacks any kind of explanatory or heuristic power. Rather, "the important thing is to understand the internal mechanism that brings the preformed seminal parts together in such a way that an organism results which possesses the same specific structure its ancestor possessed"[27]; this is where he turns to Geoffroy's theory of chemical affinities as an analogy for this internal mechanism. Maupertuis stipulates that the mixture of the seminal liquids of organisms is accounted for by special attractive forces similar to those that explain the mixture between reactive chemical substances.

Geoffroy's concept of affinity is discussed in greater detail in Chapter 7. However, for our purposes here, it will suffice to say that Geoffroy considers chemical affinities as the bonding forces present in bodies and substances, the actions of which are differentiated in a qualitative fashion. Unlike forces of universal attraction, chemical affinities are conceived as capable of acting selectively, in

---

[24] Maupertuis, *Œuvres*, 146–147.
[25] Ibid., 158.
[26] Carvalho Ramos, "Organic Monadology in Maupertuis," 18.
[27] Ibid., 19.

accordance with the substance involved. . . . The figure of chemical molecules corresponds to the form of the seminal parts. . . . [In] the sphere of organic generation, a modulating effect that is an even stronger determinant may be expected, for the figure, but for the form of the seminal parts.[28]

It is clear, however, that in attributing "memory" to the seminal matter involved in epigenesis, Maupertuis is endowing the seeds themselves with life and with a type of consciousness, so that the seeds contain within them not merely a material form but also a representation of that form, contained as a memory. It is at this point, then, that Maupertuis analogizes his theory to Leibniz's monadological model and to its implicit theory of representation. He states, "if we wish to be able to formulate anything about this, even if we can only conceive it through analogy, we must have recourse to a principle of intelligence, to something akin to what we call *desire, aversion, memory*."[29]

There is a sense in which, for Maupertuis, the seeds of life are what may be called "living *minima*," although he does not explicitly use this term. In fact, and as Cassirer reminds us, Maupertuis resists the temptation to describe atoms as being alive since "Life is a property of corporeal entities and conversely, *minima* are not alive. For Maupertuis, it is best to speak of molecules, that is, of composites of atoms that are "endowed with desire, aversion and memory."[30] Yet, as Diderot inquires in his debate with Maupertuis over precisely this issue, should the "endowment" or the "attribution" of properties "be applied to the element [atom] or the organizational whole?"[31] The staunch reductionist La Mettrie was himself profoundly aware of the implications for the future of Maupertuis' proposal regarding materialism. In *L'Homme Machine*, La Mettrie states that the Leibnizians have "with their Monads . . . spiritualized matter rather than materialized the soul,"[32] and he accuses Maupertuis of engaging in the same kind of spiritualism inspired by Leibniz, but of hiding this under the pretense of speaking of "endowed" molecules.

Although we do not agree with La Mettrie's reductionism, one cannot fault him for remaining skeptical about Maupertuis's denial of spiritualist accusations, especially since he never clearly explains what he means by the term *endowed* beyond repeating that "God did endow the smallest parts of matter *like* ('semblable') what we refer to as 'desire, aversion and memory' in us."[33] We grant that, for Maupertuis, molecules possess "positional" memory, rather than conscious memory, and that he does not claim that molecules have the same sort

[28] Ibid.
[29] Maupertuis, *Système de la nature: Essai sur la formation des corps organisés*, 147.
[30] Ibid., 173.
[31] Wolfe, "Endowed Molecules and Emergent Organization: The Maupertuis-Diderot Debate," 42.
[32] La Mettrie, *L'Homme Machine*, 149.
[33] Wolfe, "Endowed Molecules and Emergent Organization: The Maupertuis-Diderot Debate," 49.

of intelligence possessed by human beings. Additionally, when he claims that molecules are endowed with a drive, desire, or instinct to "combine with other molecules in structurally coherent ways," he is indeed speaking metaphorically. As already pointed out, the "drive" or "desires" to which he refers are analogical to chemical affinities rather than to psychological states. Nevertheless, Maupertuis does also postulate molecules that are endowed with higher-level dispositional and relational properties, while failing to explain how molecules could have such properties, limiting himself simply to stating that they were thusly endowed by God.

Philosophically, the appeal to God as a heuristic principle fails to explain anything, particularly because God's existence and role in creation are simply assumed and never established. Rather than availing himself of a structural explanation that might better account for the emergence of higher-level properties such as desire and memory, as Diderot chose to do, Maupertuis simply "spiritualizes" the molecules. To his credit, however, Maupertuis likely chose this Leibnizian approach because he was fully aware that Diderot's solution was also not quite satisfactory. After all, to the extent that Diderot was not a Newtonian and did not embrace notions of affinity or forces of attraction and repulsion, he could not properly defend his conclusion that simple atomic particles of inert and mechanistic matter could, by simple aggregation of increasing complexity, result in material bodies that displayed higher-level properties. Thus, in spite of its regression to a form of vitalism that was unacceptable within the materialist model, Maupertuis's proposal was a genuine attempt at grappling with the heuristic difficulties of mechanistic materialism.

Faced with the choice between Cartesian dualism, reductive materialist monism, or a structuralist position that made no sense from the mechanistic perspective, Maupertuis chose to take the road less traveled, although it would inevitably lead to an explanatory dead end. Yet, his failure merely emphasizes the impossibility of fully explaining the existence of life and of consciousness without appealing to emergent properties, though such a theoretical move was incoherent from a mechanistic materialist perspective, in spite of Diderot's best efforts. If, to explain the properties of macroscopic objects and of more complex living systems, one must assume that these properties are present in the atoms or molecules that constitute such objects, then it is clear that such a reductionist conception of reality must inevitably eradicate the concept of emergent properties of wholes, since these cannot be obtained either from the singular parts or from their summative aggregation. Within this very limited theoretical framework, one cannot explain how one could produce life and consciousness by aggregating inert and inanimate singular particles. Thus, one must resort, as does Maupertuis, to endowing the molecules themselves with some degree of life and awareness. Yet, this strategy itself is unacceptable within the ontological framework within which Maupertuis himself was operating.

Ironically, Maupertuis's position is in some sense just as reductionistic as the mechanistic materialism of La Mettrie. Because he is unable to explain complexity or the emergence of higher-level properties within the restrictive framework of early modern mechanistic atomism, Maupertuis is forced to postulate that lower-level entities must have some form of life. But to require that the properties displayed at higher ontological levels also be present at lower ontological levels is nothing if not eminently reductionistic. We are indeed much more familiar with the idea that reductionism is a flaw of mechanistic and eliminative materialism, which would have us deny that anything such as life or consciousness could exist if lower-level entities do not display these properties. However, although Maupertuis's position is not eliminative in this sense because it seeks to explain the possibility of life and consciousness, it is still reductionistic because it can only explain these properties by endowing them to fundamental particles.

In the chapter that follows, we shall examine in detail the early modern mechanistic philosophy that we have only discussed briefly in this chapter, highlighting the limitations of Cartesian mechanism, discussing Pierre Gassendi's conception of molecule, and emphasizing Robert Boyle's attempt to reconcile the mechanical philosophy with chemical practice by formulating an early version of structuralism and by suggesting an emergentist view of chemical properties. We shall see that Boyle's attempts to conceptualize microstructure were ultimately destined for failure precisely because, without a concept of bonding forces, he was unable to account for the stability of structured aggregate of fundamental particles.

2

Early Modern Mechanistic Atomism and the Concept of Structure

The Mechanistic Theory of Matter

From a philosophical perspective, one of the most significant developments that helped usher in early modern science was the mechanistic theory of matter or mechanicism, as it was advocated by 17th- and 18th-century philosophers and scientists such as René Descartes, Baruch Spinoza, and Robert Boyle. What distinguishes mechanicism from its Aristotelian and vitalistic predecessors is its attempt to explain natural phenomena by reducing all higher-level qualities of material bodies to the mechanistic properties of fundamental particles. A correlate of this reductionist perspective was the view that matter is an inert substance devoid of any "force," inherent motion, or self-organization. In this view, all phenomena and interactions in nature are produced by the impact and collision of particles, which are themselves governed by the fundamental laws of mechanics.

Although, as established in Chapter 1, the particulate theory of matter was compatible with a vitalistic cosmology, the idea that all material things are made up of infinitesimally small and invisible particles or corpuscles was especially suited to the mechanical philosophy. Some, though not all, corpuscularians believed that fundamental particles were absolutely simple substances that could not be further divided. These simple substances were called "atoms," a term deriving from the Greek *atomos* meaning indivisible. The term was adopted as part of the modern revival of the ancient atomism advocated by Democritus, Leucippus, Epicurus, and Lucretius. According to atomists, the only properties of atoms are their shape, size, mass, spatial position, and state of motion or rest. All of these properties were regarded as mechanistic and were considered quantifiable at least in principle, since each of them could be measured given the proper technology and method. Although atoms were considered to represent the terminus of all material bodies, they were neither observable nor detectable by early modern technologies or experimental methods. This meant that, from the 17th to the 19th centuries, the atomistic theory retained the status of being purely metaphysical speculation, especially from the point of view of pragmatic-minded experimental chemists.

From the Atom to Living Systems. Marina Paola Banchetti-Robino and Giovanni Villani, Oxford University Press.
© Oxford University Press 2023. DOI: 10.1093/oso/9780197598900.003.0003

In spite of its speculative status, one can confidently state that it is the revival of ancient atomism in the 17th century that spurred the development of early modern science. According to early modern atomists,

> the physical world is represented by particles of matter in motion and can be interpreted by the laws of motion determined by statistics . . . dynamics [and] mechanics. . . . Natural phenomena such as air resistance, friction, the different behaviors of individual bodies, the qualitative features of the physical world were now considered irrelevant to the discourse of natural philosophy or viewed as disturbing circumstances which were not . . . to be taken into account in an explanation of the physical world.[1]

This view is very similar to Galileo's claims about the nature of physical properties. For mechanistic atomists, "any explanation of natural events requires the building of a mechanical model as a 'substitute' for the actual phenomena being studied."[2] Mechanistic explanations, which rejected any reference to vital forces or final causes, provided an alternative to the vitalistic and teleological models that had dominated natural philosophy through the Middle Ages, into the Renaissance, and up to the end of the 16th century.[3] To gain a better understanding of early modern atomism, it is important to examine the ideas of Pierre Gassendi, whose revisions of ancient atomism were responsible for the diffusion of this particulate theory of matter in early modern Europe.

Prior to discussing Gassendi, however, it is important to note that atomism is one among several corpuscularian theories, not all of which accept the existence of atoms as indivisible particles. One such nonatomistic view is Cartesian corpuscularianism, which must be examined first in order to understand why Gassendi chose to break with this conception of matter. We shall begin by addressing Descartes' rejection of the concept of substantial form, since this issue is deeply connected to both his mechanical philosophy and his nonatomistic corpuscularianism.

The Cartesian Mechanical Philosophy and the Denial of Substantial Form

One of the most important advocates of the mechanical philosophy is the 17th-century French philosopher and mathematician René Descartes (1596–1650).

---

[1] Rossi, *The Birth of Modern Science*, 122.
[2] Ibid., 125.
[3] Ibid.

Although he himself embraced a rationalistic perspective, his theory of matter also found strong support among many of the leading empiricists of his time. Whether in its Cartesian or in its empiricist iteration, the mechanical philosophy rejected the Scholastic theory of substantial form, with the two chief criticisms being that the notion of substantial form was obscure and that forms should not be treated as substances.[4] Despite their many philosophical disagreements, Cartesian and empiricist mechanicists were of one mind that the concept of substantial form, understood as something over and above the material constitution of bodies that served as the source of their phenomenal properties, had absolutely no heuristic or explanatory value.

Descartes interpreted substantial forms as something that is subsistent in material bodies. In a letter to the Dutch philosopher and physician Henricus Regius (1598–1661), Descartes stated that "to prevent any ambiguity of expression, it must be observed that when we deny substantial forms, we mean by the expression a certain substance joined to matter, composing with it a merely corporeal whole, and which no less than or even more than matter, is a true substance or a thing subsisting per se, since it is said to be an actuality, and matter only a potentiality."[5] Descartes found the notion of substantial form to be incompatible with a truly mechanistic corpuscularian view of inanimate material bodies. In fact, he believed that the mechanical philosophy could dispose of this idea, since there is only one *res extensa* (extended substance) that is shared by all material bodies, and its mechanical affections fully account for all the phenomenal qualities that we observe. Because he believed that the qualities and behaviors of both natural bodies and material artifacts are fully explained by their mechanical affections, Descartes rejected the Scholastic distinction between natural and artificial bodies, according to which the natural body is endowed with substantial form while the artificial body lacks substantial form.

Although Descartes believed that the mechanistic hypothesis can fully explain both the properties and the behavior of all material bodies without recourse to any notion of form, substantial or otherwise, his formless mechanism did not resolve the mereological problem of how one accounts for both the synchronic and diachronic unity that distinguishes composite (i.e., divisible) wholes from mere collections or heaps. I specify composite wholes here because Descartes had no trouble accounting for the synchronic and diachronic unity of simple (i.e., indivisible) wholes such as *res cogitans* (thinking substance). In this unique case, Descartes willingly appealed to substantial form, as had his Scholastic predecessors.

---

[4] Pasnau, "Form, Substance, and Mechanism," 45.

[5] Descartes, *The Correspondence between Descartes and Henricus Regius / De briefwisseling tussen Descartes en Henricus Regius*, 106.

The difficulty for composite wholes, however, also translates into a difficulty for realism about natural kinds. Since, according to Descartes, material bodies do not have distinctive essential properties because all of their properties are reducible to and accounted for by the properties of their elementary particles, he could not account for why extended bodies could legitimately and nonarbitrarily be classified as belonging to distinctive material species.

Because Descartes, like David Hume after him, denied the presence of any sort of metaphysical "glue" that holds material bodies together, in space and time, he had to find some other explanation for unity and stability that would be consistent with the mechanical philosophy. He explained unity and stability by postulating that elementary particles have distinctive shapes that "fit" together, much like the pieces of a jigsaw puzzle do. His mechanistic commitments precluded endorsing any sort of affinity, force, or occult quality that holds particles together to account for a body's unity, nor did he wish to reconceptualize the idea of form in mechanistic terms. Thus, the mereological question of what accounts for the unity of composite wholes remains unanswered.[6] We shall see that one advantage of the systemic approach to chemistry that we advocate in this book is that it successfully addresses this question by treating such wholes not merely as composite wholes but, more importantly, as complex wholes. The difference between composite and complex wholes will be addressed in later chapters.

In his *Principia Philosophiae* (*Principles of Philosophy*, 1644), Descartes argued that particles are separated by their motions and are joined together by their state of rest. His argument suggests that appealing to another substance beyond that of the particles themselves to explain unity would lead into a regress when we try to explain how this other substance holds the particles together. On the other hand, appealing to a mode other than rest to explain unity would violate the principles of mechanism by invoking a nonmechanistic principle. Yet, the idea that composite wholes are unified simply because their particles are at rest is equally unsatisfactory because it cannot account for the stability of material bodies that warrants classifying them into natural kinds or material species. The inability to accommodate any notion of form or to reconceptualize this notion in mechanistic terms is one of the weaknesses of Descartes' mechanistic ontology. As we shall see in the next section, his mereologically simplistic rejection of substantial form and his rejection of the indivisibility of particles led to a number of disagreements with the important French corpuscularian, Pierre Gassendi.

---

[6] For a detailed discussion, see Pasnau, "Form, Substance, and Mechanism."

Pierre Gassendi and Mechanistic Atomism

Pierre Gassendi (1592–1655) was arguably the strongest proponent of mechanistic atomism in early 17th-century France. His *Philosophica Epicuri syntagma* (1649) is significant because it contributed to the acceptance of Epicurean atomism in early modern Europe by making atomism compatible with Christianity. Gassendi achieved this by arguing, against Epicurus, that motion is not inherent to matter and that an external cause is required to impress motion upon atoms. This external cause, he argued, is God.[7] Since Gassendi sought to attribute both empirical reality and causal power to atoms, he rejected Descartes' identification of atoms with mathematical points because mathematical entities have no empirical reality or causal properties. In this regard, he also rejected the Cartesian view that the essential property of matter is extension, since this would render incoherent the idea that fundamental material particles are indivisible and, thus, atomic. In his later work, *Syntagma philosophicum* (1658), Gassendi defended the atomistic view by claiming that matter is that which has dimensions and is capable of resistance, so it follows that prime matter is constituted of solid and indivisible particles.

For Gassendi, the number of distinct atoms is inconceivably large, but the forms of atoms are finite, as are the exemplars of those forms. In addition to discussing the properties of single atoms, which are size (*moles*), shape (*figura*), solidity (*soliditas*), and mass (*pondus, gravitas*), Gassendi also postulated properties that pertain to groupings of atoms. Like Daniel Sennert (1572–1637) before him, Gassendi believed that "the primordial atoms combine with one another to form compound corpuscles,"[8] which he called molecules (*moleculae*) or concretions (*concretiunculae*). The properties of molecules are the arrangement of atoms (*ordo*) and the position of each atom with respect to the other atoms (*situs*) in that arrangement. Gassendi also considered molecules to be stable aggregates that cannot be analyzed into their constituent atoms by ordinary chemical processes. For him, molecules served as the intermediaries between indivisible atoms and tangible and perceptible material bodies. Since they are produced by chemical resolution, Gassendi considered molecules to be operationally elementary, although he regarded atoms as the only true elements since they are the only particles that are completely fundamental, basic, and indivisible.

Gassendi believed that there are several intermediary levels of compounded corpuscles between the fundamental atoms and concrete empirical bodies. Thus,

---

[7] Shapin and Schaffer, *Leviathan and the Air Pump: Hobbes, Boyle, and the Experimental Life*, 202.

[8] Newman, *Atoms and Alchemy: Chymistry and the Experimental Origins of the Scientific Revolution*, 191–192.

he believed in varying degrees of molecular complexity, and it is these complex chemical molecules that compose the elements of traditional alchemy and chymistry (sulfur, mercury, salt, earth, and water). For Gassendi, the concept of complex chemical molecules whose properties account for the chemical nature of elements and reagents was compatible with his mechanistic atomism. Gassendi thought that structural alterations to molecules produce qualities in substances and that such changes can be induced by chemical operations.[9] He suggested that the molecules of chemical elements or principles characterize the various species of bodies, depending on their proportions and composition. However, he found it difficult to distinguish homogeneous bodies with identical molecules from mixed bodies, especially when determining the nature of metals.[10]

As suggested above, a significant point of contention between Descartes and Gassendi was the issue of the divisibility of fundamental particles. This issue provoked a significant amount of debate among early modern corpuscularians, many of whom took Descartes' side and wondered whether any physical particle could ever truly qualify as atomic. Descartes took this position because he thought that, if the essential property of matter is extension, then matter is infinitely divisible. Thus, although the minute particles in nature may be called atoms for the sake of convenience, they are not truly *a-tomos* because any amount of extension, no matter how small, can be subdivided even further, ad infinitum.

In his *Principia philosophiae*, Descartes states: "even if we suppose that God wished to reduce some part of matter to a minuteness so extreme that it could not be divided into smaller parts, we could not conclude from this that it is indivisible. For if God rendered this part so small that no creature could divide it, he certainly could not deprive himself of the power to divide it because it cannot be that he could diminish his own power."[11] Here, Descartes is saying that the indivisibility of elementary particles is a practical rather than a conceptual matter. In principle, all particles are divisible because they are extended. He concluded that "there cannot be any atoms or particles of bodies that are indivisible, as some philosophers have imagined. No matter how small we imagine particles to be, since they must be extended, we cannot conceive that there is not one among these that cannot be further divided in two or more even smaller particles. . . . We shall say then that the smallest possible extended particle that can exist can always be divided."[12] Thus, for Descartes, no physical particle could ever be truly atomic, and the only true atom was the dimensionless mathematical point.

As noted earlier, Gassendi rejected the Cartesian idea that extension is the essential property of matter and, consequently, that any amount of physical matter

---

[9] Ibid., 192.
[10] Pinet, "La philosophie de la matière de Galilée à Newton," 67–82.
[11] Descartes, *Principia philosophiae*, 51.
[12] Descartes, *Les principes de la philosophie*, 74.

can be divided ad infinitum. For Gassendi, fundamental particles are solid, impenetrable, mereologically simple, and indivisible. However, what this debate between Descartes and Gassendi demonstrates is the intimate relationship between metaphysics and natural philosophy in the 17th and early 18th centuries. The severing of this relationship was an important motivation for Antoine Lavoisier's reconceptualization of chemical theory and practice in the late 18th century by focusing on identifying elementary substances rather than speculating about the divisibility or indivisibility of fundamental particles.

This discussion points to a persistent tension that existed in philosophy between two different conceptions of fundamentality: the metaphysically fundamental versus the empirically fundamental. That which is metaphysically fundamental is the most basic substance in nature, while that which is empirically fundamental is what cannot be further analyzed by the chemical procedures available at a given moment in history. These would be the homogeneous substances that remain when compounds are analyzed and cannot be further broken down into simpler substances. Centuries later, the German chemist Friedrich Paneth (1887–1958) highlighted the tension between these two conceptions of fundamentality in his discussion of the dual meaning of element in chemistry. According to Paneth, the concept of *atom* as espoused by ancient and early modern atomists pointed to what he called basic substances (*Grundstoffe*) as metaphysically fundamental entities, while the concept of *element* as espoused by Aristotle and by practicing chemists pointed to simple substances (*einfacher Stoffe*) as empirically identifiable constituents of homogeneous compounds.[13]

We will return to this point repeatedly since, as Paneth also demonstrated, the tension between these two understandings of elementarity and the debates that ensued therefrom are not only at the heart of early modern science and ontology but also helped to shape the development of chemistry from the 18th to 19th centuries and well into the 20th and the advent of quantization theory. Paneth also underscored that naïve realism about basic substances had played a significant role in attempts to reduce chemistry to physics, which would ultimately undermine the heuristic status of chemistry as an independent science.

As the later chapters of this book demonstrate, such naïve realism must be overcome if we are to understand the journey of modern science from the early modern atomistic perspective to the 19th-century conception of chemical atoms, structural chemistry, complexity theory, and the theory of emergence. In particular, such naïve realism must be overcome if we are to eliminate its intrinsic reductionism, that is, the view that all phenomena and processes in nature must be

---

[13] Paneth, The Epistemological Status of the Concept of Element (I)," 1–4 and "The Epistemological Status of the Concept of Element (II)," 144–160.

explained in terms of the properties of their putatively most basic constituents and, instead, embrace the contemporary scientific perspective that recognizes the complex nature of the relation between lower- and higher-level properties.

Gassendi also took issue with Descartes' complete rejection of form since this would be equivalent to denying substantial identity. Gassendi pointed out that, when material bodies undergo changes, something sustains those changes but also endures in time. However, this does not mean endorsing the Scholastic conception of substantial form, which was an idea that all mechanistic corpuscularians rejected. In fact, he did not purport to answer the question of what it is that endures while sustaining change, that is, the question of what the subject of change is. He simply limited himself to pointing out that Descartes' ontology was incomplete in this regard. Gassendi claimed that, although we may never be able to answer the question of what the subject of change is and although our ontology may always remain incomplete, we should at least admit this failure rather than pretend there is no problem, as Descartes seemed inclined to do.

Another disconcerting issue that Gassendi detected in Cartesian corpuscularianism was its failure to explain an important mereological "element in all natural development, for completed wholes [attain] to natures that their parts [do] not possess."[14] Like Sennert and Boyle, Gassendi opted for the hypothesis that "the primordial atoms combine with one another to form compound corpuscles,"[15] which he called molecules (*moleculae*). These molecules are mereologically distinct from atoms in that they are composite and stable wholes that possess qualities not possessed by their constituent atoms. To properly address the question of whether wholes acquire properties that are "over and above" those of their discrete parts, one must engage in a detailed discussion of mereology and, in particular, of chemical mereology. Unfortunately, Gassendi offered no mereological explanation for the stability of molecules or for the relation between molecules and their constituent particles. He did, however, admit that this type of explanation would be the only satisfying alternative to the reduction of higher-level properties to the mechanistic properties of fundamental particles.

Although, as an atomist, Gassendi believed in the ontological dependence of the higher-level properties of molecules on the lower-level mechanistic properties of atoms, he also recognized that this ontological dependence did not entail explanatory reductionism. Such reductionism had little heuristic value for Gassendi, at least when one is confronted with the task of explaining the stability and reactive properties of chemical substances. Because molecules are produced

[14] Gregory, "The Animate and Mechanical Models of Reality," 311.
[15] Newman, *Atoms and Alchemy: Chymistry and the Experimental Origins of the Scientific Revolution*, 191–192.

by chemical resolution, for Gassendi they were in a certain sense elementary, although they are not simple or fundamental particles as such. Although Gassendi did not speculate about what it is that makes molecules stable, that is, what holds them together, he did hypothesize that molecules are compound corpuscles that cannot be further analyzed and that serve as the intermediaries between indivisible atoms and tangible perceptible properties. This, however, is the maximum position that he could reach within the limits of the strictly mechanistic point of view. This limited position would be overcome only at the historical point when chemists would finally postulate that atoms could have distinctive qualities and that they could thus form bonds with each other. What made this change possible will be discussed in later chapters.

In the next section, we shall see that Gassendi's ideas had a profound impact on the development of English science and particularly on the work of Robert Boyle. Although Boyle did not avail himself of the concept of atom and preferred to remain a corpuscularian, the notion of compound wholes developed by Gassendi inspired Boyle's version of chemical structure as a reconceptualization of the idea of form in mechanistic terms. Boyle's distinctive structuralism would thus address many of the concerns Gassendi raised against the Cartesian rejection of substantial form.

Robert Boyle and Form as Mechanistic Structure

Mechanistic atomism successfully made its way to England due to Gassendi's influence on Walter Charleton (1619–1707). Charleton's work served as the primary vehicle for the acceptance of mechanistic atomism in England, where it exerted considerable influence on Robert Boyle (1627–1691). Boyle is one of the first English scientists to embrace the so-called purified Epicurean atomism of Gassendi and Charleton, although he adjusted Epicurean atomism to render it compatible with the Cartesian corpuscularian view that he preferred. Gassendi's revision of Epicurean atomism became particularly attractive for later natural philosophers such as Boyle because of his insistence that nature should not be regarded as a causal agent, since nature was considered to be devoid of purpose, volition, and sentience. For Gassendi, material nature was inanimate, and the only and ultimate source of motion and agency was God.[16] However, one of Boyle's primary heuristic reasons for favoring Gassendi's mechanistic atomism over earlier theories of matter (such as the Scholastic theory of substantial form or the Paracelsian theory of the *tria prima*) was that these theories could not

---

[16] Shapin and Schaffer, *Leviathan and the Air Pump: Hobbes, Boyle, and the Experimental Life*, 202.

properly explain the chemical reactions and qualitative changes observed in the chemical laboratory.

By the mid-17th century, it was evident that the active role of salt, sulfur, or mercury in some form could not satisfactorily explain chemical reactions or changes and that the theory of substantial form could not adequately account for chemical stability or chemical transformations in the context of analyses and syntheses, since such explanations were bound to end in circularity. This is because, if, on the one hand, one claims that analysis destroys the substantial form of a substance and that synthesis restores this form, one still has not explained what this form is or how it can be destroyed and restored. If, on the other hand, substantial form is regarded (as it was by some Scholastic philosophers) as something nonphysical that disappears upon analysis and reappears upon synthesis, then one has not explained how a chemical procedure could possibly affect a nonphysical "form." Finally, if substantial form simply means the substance itself, then we are saying that the substance has been altered because the substance has been altered, which is simply a circular explanation. To the extent that the mechanistic particulate theory of matter promised to be a more satisfactory explanatory model for changes in nature, Boyle sought to bring this theory under the compass of experiment. And to the extent that his experimental work was considered qualitatively and quantitatively superior to that of his predecessors, he exerted a significant and positive influence on the Royal Society's acceptance of mechanistic corpuscularianism.

Boyle actually sought to retain the concept of form and to reconceptualize it in strictly mechanistic and corpuscularian terms, without any reference to the notion of substantial form as understood by Scholastics and alchemists. Although the later chapters of this book will argue that chemical structure and chemical form are two very distinct ideas in contemporary chemistry, it is important to note that early modern mechanistic chemists such as Boyle conflated the concepts of form and structure. In order to understand how Boyle accomplished this task of reconceptualization, it is important to elucidate his theory of matter as he explained it in *The Origin of Forms and Qualities* (1666). This work contains the clearest statement on the subject and takes a position that mediates between the atomism of Gassendi and the corpuscularianism of Descartes.

Against both Gassendi and Charleton and in agreement with Descartes, Boyle stated that material corpuscles are, at least in principle, infinitely divisible. However, despite their theoretical divisibility, Boyle agreed with Gassendi that material corpuscles are impenetrable and indestructible by natural means. Boyle described corpuscles as *minima naturalia* and specifically stated that they are indivisible by nature, although they are mentally and divinely divisible. Thus, because the term *atom* implies absolute indivisibility, Boyle favored the term *corpuscles* when referring to the most basic particle of matter. Furthermore, against Descartes' claim that extension is the only essential property of material substance, Boyle maintained that matter conserves both shape and size and cannot be reduced to purely geometrical extension. Like visible bodies, insensible corpuscles have the three essential properties of shape, size, and motion or rest. However, once again in opposition to Gassendi and Charleton, Boyle believed that fundamental corpuscles are endowed by God with internal energy or "motive virtue." He believed that God furnished these corpuscles with various motions and directed their various movements and compositions to form the variety of inanimate and animate bodies that exist.

In *The Origin of Forms and Qualities*, Boyle claimed that fundamental corpuscles form clusters or concretions of various sorts and hierarchical levels, and, depending on the structural arrangement and complexity of such clusters, they affect the senses in various ways. This is simply another way of saying that the distinctive sensible properties of material bodies depend on the structure and complexity of the corpuscular concretions of which these bodies are made. Although Boyle did not adopt the Gassendian term *moleculae* for these corpuscular concretions, he did avail himself of the concept that is attached to this term. Thus, for him, corpuscular concretions were stable and semipermanent wholes that cannot be further analyzed by ordinary chemical analysis. That is to say, they resist the corrosive action of *aqua fortis* (nitric acid), *aqua regia* (mixture of nitric and hydrochloric acid), and fire, which were the most powerful analytical reagents of his time. When combined, corpuscular concretions form primary mixtures that, in turn, can be further combined to form different degrees of mixture. To the extent that corpuscular concretions are stable and homogeneous with regard to their structure and essential properties and that they cannot be further analyzed through ordinary chemical procedures, they can be considered to be chemically elementary. Contemporary Boyle scholars often refer to these corpuscular concretions as chymical atoms, as opposed to physical atoms, to emphasize precisely their chemically 'elementary' nature.

As Marie Boas Hall has pointed out, the concept of element has changed radically throughout the centuries[17] and more particularly so between the 16th and 18th centuries. Boyle's work signals precisely such a change in how elementary substances were conceptualized, and also it somewhat captures the distinction Paneth made, centuries later, between an element as a basic substance (physical atom) and an element as a simple substance (chymical atom or chemically unanalyzable concretion). To the extent that Boyle accepted the conventional 17th-century definition of elements as "substances necessarily present in all bodies,"[18] he rejected the existence of such substances and recognized only "corpuscles [as

[17] Boas Hall, *John Dalton and the Progress of Science*, 21.
[18] Ibid., 27.

the minutest portion of matter present in] all bodies, including those [bodies] regarded by others as elementary."[19]

For Boyle, what other chemists call elements are actually concretions of primary and fundamental corpuscles. That is, to use Paneth's terminology, they are concretions of basic substances. For Boyle, since "chymical atoms" retain their deep microstructure through chemical analysis, they remain as the final product of such analysis, and this is also what makes chymical atoms elementary.[20] This has been called an operational conception of elementary substance because it identifies the elementarity of a substance on the basis of the kinds of chemical operations that have produced it and that can be performed upon it.

This operational conception of elementary substance shows that Boyle was moving beyond Gassendi, Charleton, and Descartes by transcending mere philosophical speculation about 'basic substances.' Instead, he attempted to link claims about microstructure, which he called 'essential form,' to empirical observation by devising experiments that would isolate simple substances. These chemically elementary substances, as he conceived them, were homogeneously composed of chymical atoms with the self-same structure. His famous experiment of reduction to the pristine state, which he inherited from his predecessor, Daniel Sennert, is an example of one such experimental attempt to isolate and identify a simple substance.[21]

In spite of its prescience, Boyle's conception of microstructure suffers from endemic problems that cannot be resolved within the framework of the mechanical philosophy. More specifically, as an advocate of mechanism, Boyle rejects the concept of force because this idea was still closely associated with the vitalistic paradigms of his alchemical predecessors. For this reason, he cannot appeal to any notion of chemical affinity or bonding to explain what holds corpuscular concretions together or why microstructures retain their stability even under the most corrosive analytical conditions. Instead, Boyle conceived of the microstructure of chymical atoms in strictly geometrical terms and explained its stability in terms of the close proximity of the fundamental particles within the structure. Thus, the fundamental particles are so spatially close that even air cannot pass between them and corrosive agents, including fire, cannot pull them apart.

This explanation is completely unsatisfactory, however, because even the closest spatial proximity cannot account for why deep structure would persist even under the most corrosive reagents available. Without the concept of affinity or of chemical bonding forces, concepts that were anathema to strict mechanism, the stability of deep structure and its resistance to chemical analysis remain unexplained. Thus,

[19] Ibid.

[20] Anstey, "Essences and Kinds," 21.

[21] Banchetti-Robino, "The Ontological Function of First-Order and Second-Order Corpuscles in the Chemical Philosophy of Robert Boyle: The Redintegration of Potassium Nitrate."

Boyle failed to account for the operational irreducibility and chemical elementarity of corpuscular concretions or "simple substances." Additionally, given the lack of imaging technologies, any potential links that could be forged between microstructure and experimental observations would have been tenuous at best and would not have warranted the assumption that either atoms or fundamental corpuscles were involved at the micro level. Thus, the atomism and corpuscularianism of the 17th and 18th centuries remained primarily speculative attempts to conceptualize fundamental matter and to explain experimental observations under the umbrella of the mechanistic paradigm.

In spite of its failures, Boyle's work was an early attempt to account for what today we call emergent chemical properties, albeit within the framework of the mechanical philosophy.[22] Boyle applied to the structure of matter the same analysis that Galileo had applied to mechanics, while taking into consideration observed substantial changes. For Boyle, substantial modifications did not involve active qualities such as heat or vital force, nor did they involve a mixing of material substance and forms. Instead, Boyle held that the causal agent of change was intelligible and physical, once one accepted that one part of matter could act upon another by virtue of local motion. He accepted a basic tenet of mechanism and considered matter and motion to be universal principles of all material bodies, together functioning as the causal explanation of all substantial changes.

Unfortunately, however, Boyle's formalization of the corpuscular mechanistic philosophy was never anything more than an empirically unsupported and unsupportable inference regarding the fundamental nature of matter. It is not until we reach the time of John Dalton (1766–1844), over 100 years after Boyle, that the corpuscular philosophy could finally assume the positive significance that derives from measurement and quantification. Thus, in Boyle as in Democritus and throughout the 17th century, atomism constituted merely the precondition for an objective science rather than the discovery of an experimental science.[23]

In Chapter 6, we shall also see that, although the Daltonian contribution to the development of chemical atomism was an important step in the right direction, the mathematization that it brought to chemistry would not suffice for modernization of this science. This is because, although Dalton postulated atoms with distinctive weights, he was never able to explain the reasons for these differences in mass. Therefore, as later chapters will demonstrate, what was needed above all else was the ultimate overcoming of the mechanistic and atomistic perspective that would eventually occur in the 20th century, with the discovery of subatomic particles and the development of the systemic approach.

---

[22] Banchetti-Robino, "The Relevance of Boyle's Chemical Philosophy for Contemporary Philosophy of Chemistry."
[23] Banchetti-Robino, *The Chemical Philosophy of Robert Boyle: Mechanism, Chymical Atoms, and Emergence.*

3

Newton and the Newtonians

Newton's Relation to Cartesian Rationalism and Experimental Empiricism

The concept of matter as envisioned by Isaac Newton (1643–1727) strikes a middle ground between the Cartesian view and the view defended by Boyle. Although most of Newton's writings did not articulate his theory of matter, his views on this topic are made quite clear in his short essay "De aere et aethere" ("On Air and the Ether"), which many scholars date to 1673. In this text, Newton identifies three different approaches to the study of natural phenomena in general and of material bodies in particular. These approaches are the Cartesian Rationalistic and speculative approach, the Boylean experimental approach, and the phenomenalist approach. Although his writings clearly reflect the influence of both Cartesian corpuscularianism and Boylean experimentalism, Newton opts for phenomenalism after indicating both his agreements and his disagreements with the other two perspectives. Above all, he explicitly underscores the methodological difference that separates him from his predecessors, since he disagrees with Descartes regarding the necessity of metaphysical principles and the postulation of speculative hypotheses. He also disagrees with Boyle, however, since he considers that an experimentalism that is not grounded on solid mathematical principles can only describe but never explain the phenomena under investigation. For Newton, as for Galileo, experiment must be grounded in mathematics in order to yield a properly explanatory science.

In "De aere et aethere," Newton demonstrates his knowledge of the Cartesian philosophy as well as his qualified agreement with mechanism. Newton accepts corpuscularianism, as well as the Cartesian thesis of material homogeneity. But he does not endorse the idea that the essential property of matter is extension, as per Descartes' geometric perspective. Newton argues that, whether they be liquid, mineral, vegetable, or animal, all material bodies can be divided into parts, that the finest particles of matter are made of air, and that air itself can be subdivided into smaller parts until one arrives at the ether.

Regarding material homogeneity, Newton concludes that "]t]he matter of all things is one and the same and it transforms in innumerable ways through the

From the Atom to Living Systems. Marina Paola Banchetti-Robino and Giovanni Villani, Oxford University Press.
© Oxford University Press 2023. DOI: 10.1093/oso/9780197598900.003.0004

operations of nature."[1] If matter is one, in opposition to traditional substantial pluralism, priority must be given to the diverse operations or processes through which we can explain the variety of bodies that we perceive, that is, phenomenal diversity. Regarding corpuscularianism, he claims that "[a]ll material bodies, including the air that is derived from them, are composed of particles."[2]

Nevertheless, his agreement with Cartesianism goes no further than this. Thus, for example, according to Descartes's physicogeometric perspective, the properties of the parts are entirely reducible to their figure and their size, and their actions are entirely limited to mechanical collisions. On the other hand, according to Newton, the properties of bodies are much more specific and complex, and their operations surpass the very limited options provided by Cartesian mechanicism. Although from the perspective of physics, Newton postulates the three universal laws of thermodynamics, he moves away from Cartesianism in his approach to chemistry by postulating that atoms have an internal structure and by postulating properties that pertain inherently to substances. Thus, against the Cartesian characterization of the parts of air as spheres of medium size and velocity, Newton postulates that "the triple nature of air results from the triple nature of substances. The permanent and heavy air of metal, the exhalation of terrestrial parts such as vegetable substances, and the vapor of liquids."[3]

Newton also rejects the Cartesian proposal that the collision of parts constitutes a satisfactory explanation for change and for phenomenal variety, since he considers that different natural processes result in the modification of material bodies so that action is no longer regarded as being simply mechanistic. Newton claims that "aerial substances are very different in nature from the bodies that generate them. Metals, through corrosion, give off a permanent and true air; Vegetable and animal substances, through corrosion, fermentation, or incineration, give off air of short duration like a sigh; and volatile substances, rarefied by heat, give off a less permanent air than the rest, which we call vapor."[4] In addition to defending what can be considered to be a relatively modern conception of substance, Newton also discusses a series of physical and chemical processes that illustrate a path, very different from Descartes' approach, that elaborates a theory of elements as part of a speculative hypothesis grounded in his cosmological theory.

In spite of these theoretical and methodological differences from Cartesianism, however, Newton's theory of substance as presented in his *Opticks* is still somewhat speculative since he postulates experimentally inaccessible material particles, although he does not decisively embrace either atomism

[1] Newton, *Unpublished Scientific Papers of Isaac Newton*, 341.
[2] Newton, "De aere et aethere," in ibid., 189.
[3] Ibid., 226.
[4] Ibid., 130–131.

or corpuscularianism. He states: "it seems probably to me, that God in the Beginning form'd Matter in solid, massy, hard, impenetrable, moveable Particles, of such Sizes and Figures, and with such other Properties, and in such Proportion to Space, as most conduced to the End for which He form'd them."[5] Nevertheless, when addressing the ether in "De aere et aethere," Newton comes very close to adopting the terminology Descartes used in his *Principia philosophia* (*Principles of Philosophy*). Thus, Newton states: "and just as the bodies of this earth, when they are broken down into even smaller particles, through a violent actions, and transform into an even more subtle air which, if it is sufficiently subtle as to penetrate the pores of glass, crystal, and other bodies, we shall call this the subtle spirit of air or the ether."[6]

In this early text, Newton may have been trying to establish the explanatory power of mechanicism, including that of Cartesianism's hard ontological scheme, in order to determine whether it could sustain his own hypothesis of forces of attraction and repulsion. From the very beginning, his hypothesis would exclude the Cartesian proposal of completely inert material particles that either gain or lose motion exclusively through collisions. Such an explanation seemed both reductionistic and unsatisfactory to Newton because it failed precisely to account for a large variety of phenomena, not the least of which were the various types of reactions, analyses, and syntheses that he observed in his own chemical experiments.

If Newton's agreements with Descartes are only partial, the disagreements are much more significant. In fact, Newton rejects the idea that material parts are completely inert and, instead, considers that they are endowed with forces of attraction and repulsion. This is a nodal point because, once matter is no longer considered inert and is considered to have intrinsic properties, the road to the conceptualization of chemical substances with specific and distinctive properties has also been opened. Clearly, the origin and the possibility of those intrinsic properties remain to be explained, and, as later chapters will demonstrate, such an explanation must be a systemic one that takes into account both structure and complexity.

As already established, Newton also rejects Descartes' identification of matter with extension and considers that, although bodies display volumetric dimensions, space is to be distinguished from bodies. Many commentators argue that Newton seems to accept that observable phenomena are the result of the movements of particles and that such movements are themselves the result of the interaction between the forces of the particles.[7] But, of course, this implies

---

[5] Newton, *Opticks: A Treatise of the Reflexions, Refractions, Inflexions and Colours of Light*, xliii.

[6] Newton, "De aere et aethere," in *Unpublished Scientific Papers of Isaac Newton*, 227.

[7] Newton, *Unpublished Scientific Papers of Isaac Newton*, 92.

endowing the particles with forces, a theoretical postulation that was forbidden by Cartesian mechanicism. However, we cannot thereby establish that Newton was proposing a hard ontology, since it is not clear whether he considered the particles to be atoms or corpuscles. Nor did he address the question of where such particles originate or what causes particles to have the forces associated with them.

Newton seemed to have little interest in such metaphysical speculations and preferred to focus instead on the manner in which such forces operated in nature and on the description of the various processes associated with those forces. Thus, when he distinguishes various classes of air by referring to fermentation, corrosion, or incineration, he is clearly entering empirical territory, the territory of the natural histories that Bacon and Boyle had already widely traveled. This is the terrain of distinctive substances, rather than that of catholic and universal matter or generic metaphysical "substance." In "De aere et aethere," Newton states that "it is thus that each of these vehement agitations (such as friction, fermentation, ignition, and great heat) generate the aerial substance that is revealed in liquids through boiling and, the more vehement the action, the more abundant is the substance revealed."[8]

After conducting a large variety of experiments, Newton understands how metals produce permanent gases, and his interest in the ether, beyond being of speculative origin, is the consequence of the many results and effects obtained in his experimental work. Yet, only a few years after writing "De aere et aethere," Newton turned his back on any type of purely empirical experimentalism that lacks mathematical foundations. He adopted the Galilean position that either physical science is grounded in mathematical principles or the neat descriptions of the naturalists will fail to attain the status of science. Mordekai Feingold tells us that "[t]he publications of [Newton's] *Principia* marked a new phase in the relation between mathematicians and naturalists, not only because its immediate success greatly contributed to the calls for the mathematization of physics (including those who were incapable of understanding mathematics), but also because the triumph of the Newtonian approach seemed to sanction its application to other scientific domains."[9] Thus, although he inherited the experimental naturalism of Bacon and Boyle and agreed that observation and experiment were certainly preferable to mere speculation and to "feigned" hypotheses, Newton considered that these were insufficient to properly ground true scientific knowledge. Therefore, following the Galilean model, Newton took physics and all the sciences beyond both Descartes' speculative natural philosophy and Boyle's qualitative experimentalism.

---

[8] Newton, "De aere et aethere," *Unpublished Scientific Papers of Isaac Newton*, 226.
[9] Feingold, "Mathematicians and Naturalists, Sir Isaac Newton and the Royal Society," 87.

The Newtonian Reaction to Mechanistic Atomism

As has been established, although the mechanistic corpuscularian theory of matter dominated the philosophical and scientific discourse of his time, Newton rejected the Cartesian metaphysical physics and refused to systematize his own theory of matter. Although he did subscribe to a type of atomism, Newton also accepted the existence of the vacuum, which is completely rejected by the Cartesian perspective.[10] What was truly surprising, but also dismaying, to his contemporaries was the fact that Newton was willing to add active powers to material bodies, which many interpreted as a revival of a type of vitalist conception of matter. In the thirty-first Query of his *Opticks*, Newton wrote regarding the atoms that make up all material bodies:

> It seems to me farther that these Particles have not only a *Vis inertiae* . . . but also that they are moved by certain active Principles, such as is that of Gravity, and which causes Fermentation, and the Cohesion of Bodies. These Principles I consider not as occult Qualities, supposed to result from the specific Forms of Things, but general Laws of Nature, by which the things themselves are form'd; their Truth appearing to us by Phaenomena, though their Causes be not yet discover'd.[11]

Perhaps influenced by his own alchemical and chemical studies, Newton thus departed from strictly mechanistic atomism and postulated active principles that included gravitation and that were central to forming the world of observable phenomena. "He thus admitted that not everything can be explained by matter and motion alone, and that there is action that does not work by direct collision but at a distance."[12] Thus, by adding forces to particles and disengaging space from extension, Newton achieved a better understanding of movement, thereby permitting the shift from kinetics to dynamics.

In his definition of the concept of mass, Newton ultimately conflates mass with matter, referring to mass as *quantitas materiae* or, more simply, as *corpus*. In Definition I of his *Philosophiae naturalis principa mathematica*, he states that "the quantity of matter is the measure of the same, arising from its density and bulk conjointly."[13] In Newton's time, the three dimensions of physics—length, time, and density—were understood as specific weight. It was therefore natural and logical that mass would be defined with respect to density. From the

---

[10] Kargon, *Atomism in England from Hariot to Newton*.

[11] Newton, *Opticks: A Treatise of the Reflexions, Refractions, Inflexions and Colours of Light*, 20.

[12] Garber, "Physics and Foundations," 66–68.

[13] Newton, *The Principia: Mathematical Principles of Natural Philosophy* (*Philosophiae naturalis principia mathematica*), 9.

experimental point of view, density was obviously considered as the relation between the weight and volume of a body. From a theoretical perspective, however, this definition conformed to the atomistic notion according to which the fundamental property of bodies was the number of corpuscles per unit of volume. In Definition III, Newton describes the *vis insita* or *vis inertiae* as the intrinsic disposition within matter to resist change "by which everybody, as much as in it lies, endeavors to persevere in its present state, whether it be of rest, or moving uniformly forward in a right line."[14] The Newtonian conception of matter was thus dominated by two fundamental notions, *quantitas materiae* and *vis inertiae*, between which Newton postulated a proportional relation. In Book II, Proposition 6, Theorem 6 of the *Principia*, he describes several experiments through which he demonstrates that gravity or weight is proportional to *quantitas materiae*.

The Newtonian clarification of the concept of mass and its distinction from weight were essential to developments in mechanics, physics, and, more generally, modern science. In classical mechanics, mass becomes the fundamental characteristic of bodies and is indeed identified with "body." This definition of physical bodies as masses became the preeminent achievement required for the development of modern mechanics. In chemistry, this new concept of mass contributed to the decline of alchemy, in spite of Newton's own well-documented interest in this topic, and it favored the ascent of modern quantitative chemistry. For alchemists, "weight" had been considered an accidental property of matter, so that an increase in weight was not necessarily linked to an addition of matter. So for example, if lead were transformed into gold and increased in weight during this transmutation, this did not imply that the *quantitas materiae* had been altered. Thus, alchemists were convinced that, simply by adding a small amount of the *lapis philosophorum*, a few pounds of lead would transmutate into hundreds of pounds of gold. Only after Newton established the proportionality between mass and weight could the measurement of weight, as a systematic method in chemical and physical research, become scientifically grounded.

The concept of *vis inertiae* is also fundamental to Newtonian mechanics and dynamics, which dictate that there is no significant difference between the state of motion and the state of rest of a body. It is precisely the equivalence between these two states that is described in the law of inertia as the first postulate of Newtonian mechanics; that is, a body upon which no external forces are acting will continue in its state of motion or rest uniformly and indefinitely. Newtonian mechanics defines the geometric formalization of the motion of bodies independently of the forces that determine such motion, so that that the principle of inertia guarantees the integration of all equations regarding motion. The second principle, which states that force is equal to mass times acceleration, represents

---

[14] Ibid.

the fundamental law of dynamics. The third principle, which states that for every action there is an equal and opposite reaction, establishes the fundamental law of statics.

Given these principles, it became increasingly difficult for Newton to reconcile action at a distance with the atomistic conception of change. Mechanistic physics permitted only actions by way of contact and collision of particles. Yet, it was evident that gravity permitted for the action of particles in all of space and eliminated the distinction between occupied space and the vacuum, so that the action of atoms was not confined merely to those precise regions that were occupied by particles. As Prigogine and Stengers note,

> It seems that not only Descartes, Gassendi, or d'Alembert, but also Newton, believed that the collisions between solid atoms constitute the ultimate source, perhaps the only source, of changes of motion. Nevertheless, the dynamic description [presented by critics of mechanistic atomism] is opposed point by point to the description derived from the atomistic hypothesis. . . . What relation is there between that moral world, that unstable world in which the atoms unite and disunite, beings are born and dies, and the immutable world of dynamics that is held together by one unique mathematical formula?[15]

It was not the general problem of change by itself, however, that compromised Newtonian atomism once action at gravitational distance was discovered. A greater challenge was presented by the theory of the constitution of matter since, if force is proportional to acceleration, the collisions between solid bodies would imply infinite accelerations and the emergence of infinitely large forces. Without completely abandoning the notion of actions by contact or collision, Newton's solution was to introduce a model of repulsive forces that increased as the distance between atoms decreased. In *Principia*, for example, Newton established that the Ideal Gas Law will hold if one posits the existence of repulsive forces of inverse proportion to the square of the distance between particles. This approach seems to have significantly defined the focus of Newtonian atomism and its characteristic ambiguity. The laws of action at a distance involving forces of attraction and repulsion coexisted in a very precarious way with particles defined in a manner that implied interactions by way of contact or collision.

Overall, the fundamental Newtonian legacy to atomism is not so much the introduction of this or that force but rather the observation, albeit unclear at the time, that a coherent formulation of mechanics required the constitutive concept of force as something given a priori. That is, it required the concept of force as something that is not reducible to more fundamental principles.[16] Thus, as the

---

[15] Prigogine and Stengers, *La Nouvelle Alliance*, 92.
[16] Truesdell, *Essays on the History of Mechanics*, 1975.

investigation of Newtonian forces became one of the priorities of 18th-century rational mechanics, the repulsion between atoms emerged as a central component of the theory of the caloric.

This was a concern as far back as Lavoisier and, as we shall see in Chapter 5, the early 19th-century writings of John Dalton still reflect the clear influence of this concept.[17] Before taking the next step of examining the way in which later Newtonian physicists addressed the tension in Newtonian atomism, it is important to examine some of the other ways in which 18th-century philosophers responded to the limitations of Cartesian mechanistic corpuscularianism.

The Leibnizian Programme and the Kantian Physical Monadology

The Cartesian natural philosophy played a determining role in the atomistic conceptions of the 18th century, due both to Descartes' own contributions to the corpuscular hypothesis and to the reaction against the Cartesian theory of matter, especially in the work of the mathematician and philosopher Gottfried Wilhelm Leibniz (1646–1716). The basic element of Cartesian physics, that is, its goal of reducing all matter to pure extension, led Descartes to adopt a rather heterodox position regarding the canonical atomism of his time. This is because, although the Cartesian universe was a corpuscular system in which the actions of particles required only the most minimal contact, it was also a material plenum within which movement did not require a vacuum and in which one could not speak of the ultimate and indivisible atoms that had been postulated by the ancient philosopher Democritus and, more recently, by Gassendi.

Leibniz's reaction to this heterodoxy was to reject completely both the Cartesian and the atomistic perspectives. For him, the indivisibility of physical atoms is more apparent than real because no atom exists that is insurmountably hard, and no body exists that completely resists division. However, at the same time, Leibniz rejects the Cartesian identification of matter with extension and argues that neither movement nor action, neither resistance nor attraction, nor the natural laws that can be observed when bodies collide can be successfully explained through the concept of extension.[18]

For Leibniz, impenetrability is what constitutes the essential characteristic of matter, and it is this characteristic that permits bodies to have extension. For him, the relation between matter and extension is not fundamental but, rather, derives from the existence of certain forces of repulsion that emanate from the most

---

[17] Nash, *The Atomic-Molecular Theory.*
[18] Leibniz, *Discourse on Metaphysics and The Monadology.*

fundamental elements of all being. Leibniz named these fundamental elements monads, which he conceptualized as a type of metaphysical atom or unit of force. According to Leibniz, each body is composed of these nonextended monads, and it is the forces that emanate from such monads that create the extension of a body by pushing matter outwardly. Obviously, the Leibnizian notions of space and of matter also imply rejecting the vacuum, since it could be nothing other than a space free from the forces of repulsion that would become increasingly filled by surrounding matter.

What Karl Popper called Leibniz's Programme,[19] that is, the attempt to explain the extension of matter through the concept of force, was brought down from its metaphysical heights by Immanuel Kant (1724–1804) in his *Physical Monadology* (1756). Following Leibniz, Kant affirms that the impenetrability and the inertia of matter result from forces inherent to matter itself. However, Kant also argues that the existence of forces of repulsion renders unnecessary the appeal to ultimate extended constituents of matter, since it is enough that we conceive them as mere point centers of force. In this way, Kant identified atoms with Leibniz's nonextended monads and abandoned the microscopic particles of extended matter that had characterized classical and early modern atomism.

Since the explanation for impenetrability required that the repulsive forces between the point centers approach infinity when the distance between them approaches zero, Kant was forced to affirm the existence of finite distances between the point centers, all of which implied the existence of the vacuum. Thus, Kant elaborated a new model of extended substance that merged some of Newton's ideas with the dynamic concepts of Leibniz, all within a general framework in which there remained little trace of either Democritean or Gassendian atomism. In this new model, matter now consisted of a vacuum in which moved certain point centers of force with *vis inertiae*. These were nonextended particles that, through a profound displacement of meaning, would eventually be referred to as "atoms." It is with these concepts in mind that we now turn to the ideas of Rudjer Boscovich, who radically revolutionized nonmechanistic atomism in ways that neither Newton, Leibniz, nor Kant could have imagined.

Boscovich's Point Particles and the Unification of 18th-Century Physics

Rudjer Josif Boscovich (1711–1787) attempted to accomplish what was probably one of the most ambitious attempts to unify 18th-century physics. Since the

---

[19] Popper, *Quantum Theory and the Schism in Physics: From the Postscript to The Logic of Scientific Discovery*, I.

end of the 19th century, Boscovich has been regarded as one of the most distinguished representatives of the 18th-century atomistic perspective, despite the
fact that his ideas were quite distant from the atomistic views of his time. Indeed,
his ideas were later revived by James Clerk Maxwell (1831–1879)[20] in the context
of the kinetic theory of gases. This process of elaboration led, in turn, to a change
in the emphasis physicists gave to atomism and encouraged the development of
new conceptions of the structure of matter, which pioneered ideas that would be
revaluated in 20th-century atomic and molecular physics.

We begin to describe Boscovich's innovations more generally, and then we
turn to a detailed discussion of his views. Boscovich developed an original atomistic perspective by combining the Newtonian and Leibnizian views in a manner
reminiscent of, but still distinct from, Kant's proposal in the *Physical Monadology*.
Boscovich described nature as populated by a multitude of elementary particles
that he conceived, in a very Leibnizian manner, as nonextended units of force.
Their behavior was governed by Newton's laws of attraction and repulsion, but
while Newton had emphasized the force of attraction above all others, Boscovich
emphasized the importance of the repulsive force. He speculated that the force
of attraction prevailed at great distances but that, at short distances, the force of
repulsion made it impossible for atoms to touch. It was this force, he believed,
that rendered atoms impenetrable. Thus, the rigid surface of Democritean and
Gassendian atoms was replaced by an area of equilibrium between these two
types of forces.

Additionally, all these dimensionless atoms were identical, and the seeming
diversity between them was due to a difference in the number and position of
atoms, as well as to the relative distance between them. Furthermore, the dynamic nature of these atoms also nullified the concept of vacuum, which had
been an important aspect of both classical and Newtonian atomism. However, although Boscovich was influenced by Leibniz in these regards, it must be stressed
that he did not consider atoms to be Leibnizian immaterial units of force but
rather material, albeit still dimensionless, entities. We shall now examine these
ideas in greater detail.

The point of departure of Boscovich's reasoning is similar to the pre-Critical
Kantian phase to the extent that both thinkers were profoundly influenced by the
Leibnizian dynamic approach. Like the early Kant, and as was already pointed
out above, Boscovich strips monads of their variability and of their perceptual
powers, transforming them into primary "elements" that are all identical and that
lack any extension. Again, like Kant, he calls these primary elements "points"

[20] Maxwell is the mathematician and scientist responsible for the classical theory of electromagnetic radiation that described electricity, magnetism, and light as different manifestations of the same
phenomenon.

(*puncta*). This, however, is the end of the convergence between Boscovich's ideas and the pre-Critical Kantian perspective. At this point, Boscovich proceeds to elaborate a theory that is uniquely his, according to which extended matter is basically a structure of equilibrium between *puncta*. This equilibrium results from the action of ad hoc hypothetical structuring forces that emanate from the points. Boscovich's conception of matter resulted from his adaptation, for his own theoretical purposes, of the Leibnizian Law of Continuity that had been the foundation on which Leibniz had erected his infinitesimal calculus.

In his *Cum Prodiisset* (1701), Leibniz had expressed this law as follows: "In any supposed transition, ending any terminus, it is permissible to institute a general reasoning, in which the final terminus may also be included [*Proposito quocunque transitu continuo in aliquem ter- minum desinente, liceat raciocinationem communem in- stituere, qua ultimus terminus comprehendatur*]."[21] Put simply, Leibniz's Law of Continuity affirms that nothing in nature occurs abruptly. Leibniz clarified the application of this principle within his metaphysics in a 1702 letter to the mathematician Pierre Varignon by explaining that:

> it so happens that that the rules of the finite succeed in the infinite as if there were atoms (that is to say, assignable elements of nature) although there is not matter in them that is actually infinitely subdivided; and vice versa the rules of infinite succeed in the finite, as if there were infinitely small metaphysical entities, even though no such entities are required; and that the division of matter never arrives at infinitely small pieces: this is because everything is governed by reason and, if it were otherwise, there would be no science or law but this would not be in conformity with the nature of the sovereign principle.[22]

Assuming this principle, Boscovich concludes that contact between hard particles is not possible because this would imply abrupt changes in the velocity of particles, which would be in violation of the Law of Continuity. Following Newton, Boscovich also assumes that any action between pairs of *puncta* must be repulsive for short distances but becomes attractive if the centers of force are separated. In particular, if the particles become sufficiently distant, the force between them becomes ultimately equivalent to Newtonian gravitation. The originality of Boscovich's ideas consist of connecting the two asymptotic behaviors through a continuous function so that, for intermediary distances, the forces change their character for an indeterminate number of times, at times being attractive and at other times being repulsive.

---

[21] Leibniz, *Cum Prodiisset*, in *The Early Mathematical Manuscripts of Leibniz*, 147.
[22] Leibniz, *Letter to Varignon, 2 February 1702*, in *Leibnizens mathematische Schriften*, 93–94.

Once the Law of Continuity is applied to this model, this implies the existence of certain points in which the forces are annihilated, and there results a state of equilibrium that defines the natural scale at which the *punctas* are situated when they form an extended body. Accordingly, in Boscovich's theory, an object that we contemplate as something continuous in space is nothing other than a structure formed by a finite number of nonextended points, whose equilibrium is maintained by their reciprocal action. Neither the spatial extension of a body nor its mass are to be considered primary or fundamental properties but rather, are derived from the spatial structure of the *punctas* of which the body is constituted.

Lancelot Whyte indicates that the most important philosophical aspect of Boscovich's theory is his rejection of the atomistic dualism that had been associated with the existence of atoms and of the vacuum and his replacement of this view with a monistic conception of one domain of spatial relations between nonextended points.[23] Whyte adopts the same "punctual atomism," with the purpose of contrasting Boscovich's theory to the naïve atomism of Democritus, Gassendi, and even Newton. Other authors prefer to use the term *kinematic atomism* in reference to Boscovich's position, thereby highlighting the dynamic aspect of his conception of *puncta*. To be clear, the terms *dynamic* and *kinematic* here refer to Newtonian dynamics, rather than to the Leibnizian dynamicist approach.[24]

It is quite obvious that Boscovich's very unique and original theory could not, at the time of its conception, be considered as a further development of atomism but rather as a completely different hypothesis. In fact, however, Boscovich never uses the term *atom* to refer to the fundamental elements of his theory and had doubts regarding whether *puncta* could even be considered atomic, given that they had neither extension nor mass.[25]

In spite of these doubts and of the fact that his ideas were not fully understood by his contemporaries, Boscovich's theory represents a conceptual advance of the first order. For instance, one must keep in mind that, within 18th-century atomism, all hypotheses regarding action at a distance required the existence of a mechanistic medium. These hypotheses were often imbued with notions about a subtle fluid and actions by contact between particles. In contrast to these theories, the most modern aspect of Boscovich's approach is precisely his goal of unification, that is, the goal of explaining physical properties such as the impenetrability of matter, elasticity, crystallization, and propagation of sound cohesion

---

[23] Whyte, *Roger Joseph Boscovich, S.J., F.R.S., 1711–1787: Studies of His Life and Work on the 250th Anniversary of His Birth.*

[24] Sytnik-Czetwertyński, "The Philosophical Foundations of the Kinematic Atomism of Ruder Josip Boscovich," 139–155.

[25] Lancelot Law Whyte, "Introduction," in *Roger Joseph Boscovich, S.J., F.R.S., 1711–1787: Studies of His Life and Work on the 250th Anniversary of His Birth.*

through one simple principle. Unfortunately, although repulsion and attraction can certainly be conceptualized quantitatively, Boscovich's conceptualization was essentially qualitative in nature. This a problem because he never explains how the structure of *punctas* leads to the acquisition of properties, such as mass, in a way that can be tested and measured.

As already suggested, Boscovich's theory did not receive the recognition that it deserved until the late 19th century. Until then, Newtonianism continued to dominate the 18th-century atomistic landscape. Yet, in spite of efforts by Newton himself and by his followers, Newtonian atomism did not provide a satisfactory theory that could serve as the foundation for experimental predictions and, especially, for the prediction of chemical phenomena. Although Newton refined modern atomism by introducing force as the means of interaction between particles, a theoretically superior hypothesis to the one offered by Boyle, Newton still conceived of atoms as the most minute and impenetrable particles of matter. Unfortunately, this view is just as speculative as the earlier attempts to conceptualize elementary particles.

As the 18th century progressed, it became increasingly evident that speculations about "basic substances," whether such speculations were vitalistic or mechanistic, "offered little [that was] of practical utility [particularly] to chemists,"[26] and experimental scientists could easily dispense with any type of metaphysically speculative "deep" theory about the fundamental nature of matter. Chemists, already dissatisfied with the overly qualitative aspects of Paracelsian and post-Paracelsian alchemy and of vitalistic and hylomorphic conceptions of matter, became weary of embracing Boscovich's nonquantitative conception of *punctas*. It is precisely this skepticism regarding metaphysical speculations regarding the fundamental nature of matter and the causal role of qualitative properties that motivated Antoine Lavoisier's reconceptualization of chemistry as a quantitative pursuit, as well as his endorsement of a purely empirical and operational conception of elementary substances. This is the focal topic of the next chapter.

---

[26] Hendry, "Elements, Compounds, and Other Chemical Kinds," 865.

4

Lavoisier and the Quantification
of Chemistry

The ideas of Antoine Lavoisier (1743–1794) must be contextualized against the background of 18th-century skepticism regarding speculative theories about the fundamental nature of matter. By transforming chemistry into an entirely empirical pursuit that focused on producing quantifiable results, Lavoisier completely altered this science from what it had been in the 17th and early 18th centuries. He also reformed chemical nomenclature by focusing on the elementary composition of compound substances and identifying the "simple substances" from which compounds were composed. Lavoisier was, thus, able to codify a simple, coherent, and accurate compositional chemistry that modernized this science and set the stage for the groundbreaking developments of the 19th century and beyond. Before exploring Lavoisier's new system of chemistry, however, it is imperative to delve briefly into the philosophical history of the concept of element so as to better appreciate the significance of Lavoisier's contributions to our understanding of this concept.

The Concept of Element and the Fundamental Nature
of Material Reality

As Marie Boas Hall has pointed out, the term *chemical element* has a long, active, and tangled history. Although the word "element" has remained constant throughout this history, the concept attached to this word has changed radically in the course of the centuries, even within the two centuries that separate Boyle and Dalton.[1] Rather than detailing this very long history, the discussion here will limit itself to a brief outline of this idea to emphasize the point that, although the concept of element served to elucidate the fundamental nature of substance, this concept was often dissociated from the practice of experimental science. Despite this, one should note that some early modern chemists did attempt to theorize a conception of element that anticipated, in some ways, the groundbreaking work

---

[1] Boas Hall, "The History of the Concept of Element," 21.

From the Atom to Living Systems. Marina Paola Banchetti-Robino and Giovanni Villani, Oxford University Press. © Oxford University Press 2023. DOI: 10.1093/oso/9780197598900.003.0005

of Lavoisier.[2] With this in mind, let us briefly trace this concept from its classical roots to its appropriation by early modern chemists.

From the time of classical Greek cosmology to the early 18th century, "elementum" or elements were regarded as property-bearing first principles. One famous early example of a theory of elements is Empedocles' theory of the four elements, which was later revised by Aristotle. Notwithstanding some differences between the Empedoclean and Aristotelian theories, both philosophers regarded the four elements (air, earth, fire, water) and their associated properties (cold, dry, hot, wet) as the first principles for the properties of all compound material bodies. However, according to Aristotle, the four elements were neither irreducible nor compositionally fundamental. Instead, he considered amorphous prime matter as the fundamental substance, while the four elements represented the ways in which prime matter is manifested and organized in the world of existing things. Although many post-Aristotelian natural philosophers rejected his theory of the four elements, they retained the view that the properties of material bodies depend on those of constituent elements or first principles. We shall see that Lavoisier, on the other hand, rejected this idea, and we will show in later chapters how the modern systemic approach goes even further in establishing that the properties of compounds are not completely determined by those of their constituents.

Like atoms, elements were generally regarded as unobservable substances with no empirically determinable properties. For example, chemists such as Jābir ibn Ḥayyān (a.k.a., Geber),[3] Georg Ernst Stahl,[4] and Guillaume François Rouelle[5] considered that the elements or principles that formed the basic layer of their chemical ontologies were not experimentally accessible. Even when Pierre Macquer and other chemists claimed that the substances in the basic layer of their chemical ontologies were experimentally accessible, they were either unable to access these substances in practice or they were making incorrect inferences from their experimental observations.[6] One example of such an incorrect inference is Wilhelm Homberg's conclusion that his experiments on gold had identified elementary constituents.

For early modern chemists, the properties of individual material bodies were considered to be causally dependent on the nature of the elements that compose them, though not on the properties of such elements. However, the apparent

[2] Banchetti-Robino, "From Corpuscles to Elements: Chemical Ontologies from Van Helmont to Lavoisier."

[3] Newman, *Atoms and Alchemy: Chymistry and the Experimental Origins of the Scientific Revolution*.

[4] Stahl, *Fundamenta Chymiae Dogmaticae & experimentalis*.

[5] Lehman, "Mid-Eighteenth Century Chemistry in France as Seen through Student Notes from the Course of Gabriel-François Venel and Guillaume-François Rouelle."

[6] Ibid.

properties of most compounds were in fact extremely difficult to relate to those of their proposed constituents. Even before the time of Lavoisier, chemists realized that, while the properties of substances had to be related in some way to the *nature* of their constituents, they did not match the *properties* of those individual constituents.

There was a significant need to reconceptualize what was meant by element in order to make this concept consistent with experimental observations rather than with speculative philosophical theorizing. As was noted in Chapter 1, the work of Daniel Sennert provided a strong impetus for redefining the concept of element pragmatically so that this concept could become functional for chemical practice. To do so, Sennert drew an important distinction between how one conceives of fundamental matter and how one identifies diverse elementary substances through experiment.

Sennert was, significantly, one of the first early modern chemical practitioners to conceptualize elementary substances in terms of the limits attained by chemical analysis. This approach to conceptualizing elementarity focuses on what cannot be achieved by the available methods of analysis, which is why this notion of elementarity is considered a negative-empirical concept. This practice of defining elementary substances operationally as the limits of analysis can actually be traced back to the 13th-century work of Scholastic alchemists and, perhaps, even back to the writings of Aristotle.[7] However, the term *negative-empirical concept* was first coined in 1967 by David Knight[8] and was appropriated in 1970 by Arnold Thackray.[9] It was then resurrected in Bernadette Bensaude-Vincent and Isabelle Stengers's 1996 work, *Histoire de la chimie*.[10] In this important volume, Bensaude-Vincent and Stengers stress that "the idea of a negative-empirical concept is . . . a purely epistemological notion. It represents a new type of argument that locates the authority of proof not within reason but within experimental practice."[11]

Sennert's early attempt at theorizing a chemically viable conception of elementary substance was undertaken again by other 17th-century natural philosophers. One of these philosophers was Pierre Gassendi who, as was established in Chapter 2, believed that molecules were stable compounded corpuscles that could not be further analyzed and that served as tangible intermediaries between truly fundamental atoms and observable bodies. To the extent that molecules were produced by chemical resolution and constituted the endpoint

---

[7] Needham, "An Aristotelian Theory of Chemical Substance."

[8] Knight, *Atoms and Elements: A Study of Theories of Matter in England in the Sixteenth Century*.

[9] Thackray, *Atoms and Powers: An Essay on Newtonian Matter-Theory and the Development of Chemistry*.

[10] Bensaude-Vincent and Stengers, *Histoire de la Chimie*.

[11] Ibid., 51–52.

of analysis, Gassendi considered them as being elementary in the negative-empirical sense, although they are not absolutely simple particles.

As discussed in Chapter 2, Robert Boyle also availed himself of this operational and negative-empirical approach to conceptualizing elementary substances. To the extent that he accepted the conventional 17th-century definition of elements as "substances necessarily present in all bodies"[12], Boyle rejected the existence of such substances and recognized only "corpuscles [as the minutest portions of matter present in] all bodies, including those [bodies] regarded by others as elementary."[13] For him, what other chemists called elements were actually concretions of primary corpuscles that remained as the final products of chemical analysis. As Alan Rocke points out, this "operational criterion of elementarity gradually insinuated itself into the consciousness of chemists, so that by the time Lavoisier first clearly and unambiguously stated it in his classic *Traité élémentaire de chimie*, it could provoke but little controversy."[14]

Lavoisier's Empiricism and His Operational Concept of Element

Although Sennert, Gassendi, and Boyle all adopted the operational conception of elementary substances to different degrees, each of these natural philosophers was still committed to theorizing about the existence of putative, indivisible atomic particles and about the fundamental nature of matter. Furthermore, these natural philosophers continued to approach experimental practice in a primarily qualitative manner and, for this reason, neither the methods that they employed nor the results that they obtained could be used to ground a systematized and modern science. It is precisely these aspects of his predecessors' work that Lavoisier rejected and set out to reform.

To begin, Lavoisier regarded the concept of a fundamental, absolutely simple, and irreducible substance of which all bodies are made as a highly suspect metaphysical notion and he saw no epistemic or heuristic value in speculating about the grounding of material reality. His concerns regarded the still-prevalent habit among the chemists of explaining chemical phenomena by positing substances that were not experimentally accessible or observable or, at least, that were not accessible in the supposed compounds in which they were thought to be present.

An example of this practice regarded the calcination of metals and the positing of phlogiston as the principle of combustion and calcination. Phlogiston

---

[12] Thackray, *Atoms and Powers: An Essay on Newtonian Matter-Theory and the Development of Chemistry*, 27.

[13] Ibid.

[14] Rocke, *Chemical Atomism in the Nineteenth Century: From Dalton to Cannizzaro*, 4–5.

was believed by many chemists to be a colorless, odorless, "self-repulsive," extremely fine, and hard-to-measure ether that was responsible for chemical reactions. They believed that plants grew by absorbing phlogiston from the air, which was then re-released when plants were burned or otherwise consumed. What we call carbon dioxide today was believed to be phlogisticated (or fixed) air, and what we call oxygen was believed to be dephlogisticated air.

In the preface to his *Traité élémentaire de chimie* (1789), Lavoisier clearly broke with the phlogistonists and with early modern chemistry's still intimate relation to speculative metaphysics by explaining the general principle that he applied in his own chemical studies. He states: "when we first begin to undertake the study of a science, our relation to that science is analogous to that of children. . . . Just as in a child, it is ideas that are the product of sensation, it is sensation that gives birth to an idea, so it is for that individual who begins to undertake the study of the physical sciences: Ideas must only arise as a consequence, as an immediate result of, an experience or sensation."[15]

For Lavoisier, just as correct ideas can only arise from experience, a priori notions that are contrived by the imagination or by the faculty of reason unchecked can lead to serious epistemic and scientific errors. He proposed that "the only way to avoid such errors is to suppress or at least to simplify as much as possible our reasoning . . . which alone can lead us astray. . . . Convinced of these truths, I have imposed upon myself the rule of proceeding only from the known to the unknown, and of deducing no consequence that does not immediately derive from experience and observation."[16]

Based on these strictly empiricist principles, Lavoisier concludes that positing suspect entities whose existence cannot be established by observation or experiment is scientifically untenable. Because such entities possess no empirically determinable features, their postulation contributes nothing to experimental knowledge. This is why Lavoisier rejects any iteration of the particulate theory of matter, including the corpuscularianism of Descartes and the atomistic theories of Gassendi and Newton. According to Lavoisier, such speculations do not advance but rather hinder scientific knowledge about substances, their behaviors, their interactions, and their transformations. It is important to note, however, that Lavoisier's relation to Newtonian atomism was more nuanced than the above claims might suggest.

To obtain a better understanding of his complex relation to Newtonian atomism, it is important to examine some of the fundamental agreements between Lavoisier and the Newtonian Pierre-Simon Laplace (1749–1827) regarding the behavior of gases. Both Lavoisier's and Laplace's concepts of

---

[15] Lavoisier, *Traité Élémentaire de Chimie*, viii.
[16] Ibid., x–xi.

expansion, contraction, and change of state utilized Newton's indication that a gas composed of particles repelling each other by a force inversely proportional to distance would obey Boyle's law, and this agreed with experimental findings. Therefore, the *physical* side of Lavoisier's approach to chemistry was corpuscular or particulate in nature, and, to some extent, it was also Newtonian. Consequently, this suggests that Lavoisier may have implicitly conceptualized basic substances as corpuscles or particles. However, his compositional *chemistry* relates to a lack of comparable evidence about any chemical combinations between heat and substances. Lavoisier specifically cites this as the reason why no table of combinations had been developed for the caloric, which he identifies as the "cause of heat" and which indicates that, in this regard, Lavoisier was still willing to postulate putative entities and causes.

It is this attempt to reform the chaotic chemical nomenclature that had been built up ad hoc over many centuries that Lavoisier found necessary to apply his strict empiricist criteria. Lavoisier discussed this reform in the introduction to the *Traité élémentaire de chimie* as a way of establishing that the names of compound substances would reliably reflect their elementary composition.[17] The achievement of such a reformed nomenclature led Lavoisier to both propose and codify a simple, coherent, accurate, and experimentally fertile compositional chemistry that would be unmatched by the phlogiston hypothesis. However, in order to embark on such proposed reforms of the nomenclature of compound substances in terms of their elementary composition, Lavoisier was compelled to clarify the concept of element and what should be regarded as elementary substances.

Lavoisier's conception of element is in no way to be associated with the ancient notion of "bearer of properties," although it does remain associated with explanatory fundamentality and causal dependence. His reconceptualization required that "elements be isolatable, and therefore more concrete and material than the 'principles' of alchemists and the phlogistonists."[18] Lavoisier explains that "if by the name of element, we mean simple and indivisible molecules that compose bodies, it is probable that we do not know them: if, on the contrary, we attach the name of element or principle of bodies to the idea of the last point at which analysis arrives, all of the substances that we have not yet been able to decompose by any means are, for us, to be considered elements."[19]

Although, as has already been explained, this view of what is at least provisionally to be considered as an element is not new with Lavoisier, this operational conception of elements as "simple substances" that can be experimentally

[17] Hendry, "Antoine Lavoisier (1743–1794)."
[18] Hendry, "Lavoisier and Mendeleev on the Elements," 32.
[19] Lavoisier, *Traité Élémentaire de Chimie*, xii.

identified proved extremely fruitful in the context of his taxonomical and ex-perimental pursuits. Also original to Lavoisier was his ability to combine the operational and analytical approach with an open-mindedness about how many substances might actually be simple. This is still a reductionistic perspec-tive, since it involves the notion of something simple that underlies all complex bodies. However, it opens the door to a more pluralistic way of conceptualizing simple substances by admitting that these may be diverse in terms of their properties. Furthermore, although it might be tempting to point out the ap-parent similarities between Lavoisier's conception of element and Boyle's notion of unanalyzable chymical atoms, there is a fundamental distinction between these concepts. Although Boyle explains the inability to analyze such substances by speculating that their microstructure cannot be decomposed because of how closely placed the primary particles are to each other, Lavoisier abstains from theorizing about what might be occurring at the unobservable level to render "simple substances" chemically unanalyzable.

Such abstinence follows from his staunchly empiricist method. Restricting himself to the observable evidence meant refraining from inferring any rela-tionship between the identifiably elementary substances of experiment and what may or may not be occurring between the putatively fundamental particles. According to Lavoisier, there is a profound distinction between these two.[20] We shall see later in this chapter that, in reality, Lavoisier fell somewhat short of his own requirement to forego metaphysical speculations. However, at least in prin-ciple, his approach depended on the consistent use of an experimental method and quantitative approach that had not previously been systematically applied to chemistry, a method and approach guided solely by experimental results, measurements, and observations.[21]

It is also important to note that Lavoisier was aware that the limits of anal-ysis of his time were probably only temporary and that future analytical methods might succeed in further decomposing those substances that he had identified as elementary. He, therefore, admitted that the table of elements and the nomen-clature of compounds derived by his operational approach would be only provi-sional and was entirely open to revision. In fact, he stated: "we cannot assure that the substances that we regard as simple are not themselves composed of two or perhaps a greater number of principles. However, since these principles cannot be separated or, rather, since we have no means of separating them, they behave for us in the manner of simple substances, and we must not assume them to be composed until experience and observation prove otherwise."[22] Thus, although

[20] Levere, *Transforming Matter: A History of Chemistry from Alchemy to the Buckyball*, 81.
[21] Ibid., 80.
[22] Lavoisier, *Traité Élémentaire de Chimie*, xii–xiii.

the properties of compound substances are considered to be ontologically dependent on the properties of the chemical elements that compose them, the specific relationships of dependence identified by chemical analysis are themselves valid only until the elements can be further analyzed into simpler substances to reveal an even deeper level of ontological dependence.

For Lavoisier, this analytical understanding of elements served to supplant any a priori speculations about the ultimate particles or basic substances of which things are ultimately made.[23] However, and as Robin Hendry has emphasized, although this analytical, pragmatic, and operational definition provided a criterion for deciding when a substance should be regarded as elementary, it did not specify *what* the term *element* means.[24] The lack of a positive definition of this term is, of course, built into the notion of a negative-empirical concept. To delineate any further sufficient and necessary conditions for applying the term *element* to a substance, even if only provisionally, would have meant going beyond what could be experimentally established by analysis and reentering into the dubious territory of speculative metaphysics.

Lavoisier's Quantification of Chemistry

Lavoisier's system is the culmination of the experimental program that involved the investigation of chemical substance via analysis and synthesis. However, Lavoisier also considered that the chemical work of his predecessors had not kept pace with the otherwise ubiquitous mathematization of the Galilean sciences. Although astronomy had been the first of these sciences to be mathematized in the late 16th and early 17th centuries, with mechanics soon to follow, early 18th-century chemistry lingered behind in this regard. For Lavoisier, then, abandoning philosophical speculations about unobservable fundamental particles was only the first step toward transforming chemistry into a true Galilean science. To complete this transformation meant that chemical analysis must go hand in hand with quantification techniques. For him, what can be measured and built into a quantitative system is the most convincing evidence besides the raw evidence of the senses.[25]

According to Lavoisier, the properties of elements as simple substances meet both empirical and quantification criteria because they have empirically determinable features, they can be weighed using the precision balance, and their volumes can be converted into their corresponding weights. It must certainly

[23] Hendry, "Antoine Lavoisier (1743–1794)," 66.
[24] Ibid., 66.
[25] Levere, *Transforming Matter: A History of Chemistry from Alchemy to the Buckyball*, 80.

be conceded that some of Lavoisier's predecessors and contemporaries had also worked quantitatively. Henry Cavendish, for example, undertook many quantitative investigations in terms of measuring volume, which produced very accurate results. However, for Lavoisier, the role of quantification was fundamental to chemistry and to science in general.

No chemist before him had recognized the paramount importance of quantitative relations or used them to the same extent in the interpretation of chemical reactions. For Lavoisier, weighing was a way of regulating, shaping, and validating the results of chemical experiments and of supporting chemical theories. Together, the techniques of chemical analysis and of weighing provided a way to replace old ideas about elements and principles with the new concept of simple substances whose volumes and weights could be determined and verified empirically. It must also be clarified that measuring these properties of elements does not mean measuring the elements themselves. The qualitative differences between the elements remained irreducible to the measurable quantitative differences. This was and still remains an open question because, if mathematization implies eliminating qualitative aspects, then chemistry is not a mathematizable science.

For Lavoisier, however, theories based on anything other than experimental evidence and quantitative measurement had no value for chemical science, and his assessment included all the theories of the elements and of atoms that had been proposed by his predecessors. Unlike them, however, Lavoisier was also guided by the recognition that general principles underlie the changes in weight that accompany chemical changes. He also demonstrated that the total mass of the products of a chemical reaction is always the same as the total mass of the starting materials consumed in the reaction. Not only did he note this. but he realized that it revealed an underlying principle, the Law of Conservation of Matter.

He also established that heating substances does not always change their fundamental nature, and so he separated heat out of chemical composition so that, for example, water is water whether it is heated into steam or cooled down into ice. He explained the changes of state from solid to liquid to gas by postulating a new ether that he called the caloric, which was believed to penetrate a block of ice and to melt into water by pushing the ice particles apart. According to Lavoisier, the caloric reacted with heat to produce oxygen gas.

Lavoisier applied the techniques of chemical analysis and of weighing with the precision scale to identify elements such as oxygen (in 1778), hydrogen (in 1783), and silicon (in 1787). As already established, Lavoisier also reformed the nomenclature of substances. For example, inflammable air became hydrogen, Sugar of Saturn became lead acetate, and Vitriol of Venus (or Blue Vitriol) became copper sulfate. Further expanding the discussion of nomemclature would take us beyond the main topic of this chapter, but suffice it to say here that Lavoisier did not simply change the names of substances haphazardly. He did so according to a very specific logic so that the name of a substance would reveal the element or elements of which it was composed. By 1789, Lavoisier was ready to produce the list of chemical elements. He subdivided this table into four general categories on the basis of shared chemical properties, that is, acid-making elements, gas-like elements, metallic elements, and earthy elements.

Lavoisier's experimental and weighing methods were also applied to providing an alternative to the theory of phlogiston. This theory explained the changes that occurred during combustion and calcination by postulating a principle called phlogiston, which was believed to be released into the air during these processes. Phlogiston was thought to have mass, so this theory predicted that, upon combustion and calcination, the mass of a substance would decrease due to the phlogiston being released. Lavoisier offered an alternative explanation by postulating the hypothesis that, during combustion, something is taken out of the air rather than put into it. He burned phosphorus and sulfur in the air, weighed the final products, and proved that these weighed more than the original substance. Since these results contradicted the predictions made by the phlogiston theory, Lavoisier believed that he had falsified that theory. In its stead, he proposed that the weight gained was due to air being absorbed by phosphorus and sulfur during combustion. This hypothesis turned out to be correct, since something was indeed absorbed and that something was oxygen.

Precisely because Lavoisier's system of chemistry precluded inferring any connections between chemical elements and putative fundamental particles, the formerly intimate relation between metaphysics and early modern chemistry seemed to have been finally severed. However, as suggested above, Lavoisier might not really have been completely successful in liberating himself from metaphysical or a priori speculations. For a clarification of this point, we quote Friedrich Paneth at length:

> Lavoisier obtained general recognition for the principle that one should not make *a priori* statements about the number of elements, but should determine *a posteriori* how many substances we actually find in nature which cannot be further decomposed, and regard these as the constituents of the composite ones. In this way an experimentally determinable criterion was introduced into the definition of element, and the interminable and obscure discussions about the true elements brought to an end. Nevertheless I cannot agree with those who believe that anything has thereby been changed concerning the "metaphysical" nature of the concept of element. The essential point, after all, is the assumption, which is retained, that the simple substance is present in some latent form (i.e. in such form that the properties which it has in the pure state are not recognizable) in

the composite ones, and that it explains their behaviour. . . . Lavoisier allowed a place amongst the simple bodies to the chemical compounds lime, magnesia, baryta and alumina, which had not been decomposed at the time, although he considered that they were probably oxides; but he omitted potash and soda, because he was already firmly convinced that they were composed of basic constituents still to be discovered. . . . Here, then, Lavoisier himself assumed hypothetical elements; these had no more been produced than, say, the purest earth of the Aristotelians, or the special sulphur of the Spagirists (which is not the same as common sulphur).[26]

In 1781, Lavoisier confirmed that pure water was produced when inflammable air (hydrogen) was burned in the presence of pure air (oxygen). Cavendish had already obtained these same results, but he had interpreted them in accordance with the phlogiston theory. On the other hand, Lavoisier postulated that hydrogen and oxygen were chemically elementary and, thus, that water was a composite of two simple substances, rather than being itself a simple substance or a "bearer of liquidity" as had been proposed by the ancient theory of the elements. The most impressive aspect of this discovery is that the composition of water was the result of a reaction between two simple bodies that independently possessed quite different properties from those of the mixt. Liquid water was thus discovered to be a composite of gases.

Lavoisier's *Traité Élémentaire de Chimie* represented a crucial turning point not only for chemistry but also for the entire conceptual structure of modern science, precisely because this work involved postulating that reality was far more complex at the elementary level than his predecessors had ever imagined. In fact, this postulate is the prerequisite that allowed him to identify a total of 33 elements, thus taking the first and essential step that eventually led to the development of the Periodic Table of the elements and the identification of millions of chemical substances, both elementary and composite.

Because Lavoisier admitted that his table of elementary substances was only provisional, he recognized that neither experience nor experimental observation could furnish final proof that a particular substance was truly elementary. This was the turning point for the entire conceptual structure of science because it undermined the idea that scientific theories could be demonstrably true. Rather, it seems to have been Lavoisier's position that the best that can be achieved by experimental science are provisional conclusions about their objects of study. If such conclusions are quantifiable and grounded in proper experimental methods, and if they are devoid of any unjustifiable or speculative assumptions,

---

[26] Paneth, "The Epistemological Status of the Chemical Concept of Element (II)," 147.

they not only have valuable heuristic power but also contain within them the fruits for further investigation and growth.

Although Lavoisier's predecessors understood that nature is complex, they had always assumed that deep and fundamental reality was simple and monadic. With the turn taken by Lavoisier, chemists began to think of nature as being complex at every level, although the full implications of this idea would not be realized until the 20th-century development of quantum physics and quantum chemistry. Before this could happen, however, the atomic theory would have to be resurrected and revised, and the concept of atom would have to be reconceptualized. Yet, to the extent that Lavoisier's approach had severed the relationship between chemical practice and metaphysical speculation about the fundamental nature of material substance, no atomic theory would be credible unless it disentangled itself from metaphysics and proposed an empirical and quantitative concept of atom. The disentanglement and transformation of this concept finally occurred in the 19th century, with the development of chemical atomism in the work of John Dalton.

5

Affinity, Compounds, and the Laws of Definite Proportions

As discussed in the previous chapter, Antoine Lavoisier concluded that in chemical compounds "the quality and the quantity of the constituents is the same, and modifications alone can take place."[1] That is, the kind and amount of the matter in the universe remain constant. Lavoisier introduced the precise formulation of the principle that limits the possible changes that matter can undertake. Thus, many different substances may be produced by the various combinations of constituents, but the amount and the ultimate properties of these constituents cannot be altered. It is impossible to change either the quality of elementary substances or their quantity. Hence, near the end of the 18th century, chemists recognized certain limitations to their power for producing chemical changes.

The late 18th and early 19th centuries, however, also provided evidence for further limitation to chemists' work, when it was established that the combination of constituents in a chemical compound was itself fixed by certain laws of proportion. This fundamental "fixity of composition" determined that two substances that have the same properties must also contain the same constituents in the same proportion or ratio. However, Berzelius's discovery of isomerism in 1827 showed that the converse is not true; that is, two substances with the same composition in the same proportion do not necessarily have the same properties, since their properties are also a function of their structure. Quantitative analyses of compounds had been done prior to the beginning of the 19th century, and, at times, the fixity of composition may have been tacitly assumed. However, it had never been proven or formulated as a law, although it eventually had to be recognized and accounted for. Attempts to do precisely this gave rise to a considerable amount of controversy among chemists.

Another essential moment in the genesis of the concept of chemical compound as we conceive it today occurred when the number of such substances was no longer considered infinite because it was no longer strictly associated with the methodology through which chemical compounds were obtained. The late 18th- and early 19th-century work of Claude-Louis Berthollet and Joseph Louis Proust made much headway in this regard. Berthollet's theory of chemical affinities in

---

[1] Lavoisier. *Traité Élémentaire de Chimie*, 101.

From the Atom to Living Systems. Marina Paola Banchetti-Robino and Giovanni Villani, Oxford University Press.
© Oxford University Press 2023. DOI: 10.1093/oso/9780197598900.003.0006

1769 had significant implications for the theory of chemical compounds because it suggested the impossibility of an infinite number of compounds outside an interval, between a minimum and a maximum of composition.

Geoffroy and Bergman on Affinity

The causes of chemical combinations had fascinated chemists for centuries, and, in this regard, one of the prevalent ideas of the 18th century was the concept of chemical affinity, which had already been anticipated in the 13th century by Albertus Magnus (1193?–1280) to explain the forces that held substances together.

> Five hundred years later, Herman Boerhave (1668–1738) attributed chemical combinations to a special force that became greater the more two bodies were dissimilar. In the beginning, a certain number of affinities were distinguished, such as affinities of aggregation, composition, dissolution, decomposition, and precipitation. Guyton de Morveau altered the list of affinities as follows: aggregation, composition, cooperation (double), and excess. [Antoine-François de] Fourcroy later reduced the number to two—aggregation and composition (or chemical affinity).[2]

The accepted view was that substances could be ordered in accordance with their affinity to a master substance, in such a way that the substances with stronger affinity would displace those substances that had lesser affinity within the compounds in which these were present together.

As discussed in Chapter 3, Newton had suggested in his *Opticks* that the particles of bodies were endowed with attractive forces by which they interacted to produce various physical and chemical phenomena. In Query 31, Newton asked rhetorically, "Have not the small Particles of Bodies certain Powers, Virtues or Forces, by which they act at a distance, not only upon the Rays of Light for reflecting, refracting, and inflecting them, but also upon one another for producing a great part of the Phaenomena of Nature?"[3] Under the hegemony of Newtonian physics, chemists finally began to speculate that the force underlying chemical bonds was nothing other than a manifestation of the gravitational force, although they would ultimately distinguish between this chemical force and Newton's gravitational force. That is, at very short distances, the influence

---

[2] Wisniak, "Claude-Louis Berthollet," 52.
[3] Newton, *Opticks*, Query 31.

of the shape of the molecules became a decisive factor that was negligible when considering astronomical interactions.

According to the chemists, this force, which was given the name "affinity," was an essential property of matter, and it was responsible for its intrinsic activity. It was also not tied to other properties of matter, such as mass or acceleration. Finally, it was conceived as being only a force of attraction, so that instances of chemical repulsion were explained as the combination of substances with ethereal fluids. However, the Newtonian theory was still far too generic and did not account for the fact that the forces of attraction of each substance were elective and that their behavior changed under the influence of different factors, such as temperature, environment, and reagents. Because the gravitational theory did not explain the selectivity of reciprocal action on the part of substances, another explanation needed to be found.

In this context, and after reading Newton's description of affinity reactions in the *Opticks*, Étienne François Geoffroy (1672–1731) composed his first tables of affinity in 1718 and provided a systematic method of arranging both compositional knowledge and information concerning the interaction of constituent salts. In this table, the constituent salts were arranged to remind chemists that, if iron were added to a solution of silver nitrate, the iron would combine with the acid and silver would be precipitated.[4] The principle that Geoffroy used for creating his tables of affinity was to arrange similar substances so that the one following was always expelled by the one preceding, from combination with the one just heading the list.

Geoffroy describes one such experiment as follows: *Nitric Acid*: potash, ammonia, lime, zinc, lead, copper, silver. Not all chemists accepted his tabulation since, as Lavoisier had already shown, the nature of chemical combination varied with the temperature, which is something that Geoffroy did not account for. Nevertheless, his list conforms to an experiment that had already been conducted by another early contributor to the study of affinity, Georg Stahl (1659–1734), who was also the founder of the phlogiston theory of combustion. Stahl's experiment is described as follows:

> Dissolve silver in nitric acid, it will take up the silver and appear as a light liquor; into the clear and transparent liquor throw thin strips of copper foil, the nitric acid will dissolve these and will drop the silver in the form of a powder; pour this clear green solution on to lead foil, it will be attacked and the copper previously dissolved will be dropped; pour off the clear solution and pour it on to zinc, it will dissolve the zinc and allow the previously dissolved lead to drop; into this clear solution put chalk, it will be dissolved and the zin dropped; then

[4] LeGrand, *Berthollet and the Oxygen Theory of Acidity*.

to this solution add spirits of urine, which will combine with it, releasing the chalk; and finally drop in lye, the solution will take it up and allow the volatile salt to go.[5]

Through these experiments, Stahl discovered that a reaction occurring at one temperature in one direction could be reversed when conducted at another temperature. For example, at ordinary temperatures, calomel (mercury chloride, Hg_2Cl_2) is decomposed by silver, but when the temperature is raised, silver chloride is decomposed by mercury. So, when determining affinities, one must first arrange substances in the sequence of their affinities toward the one with which they all combine, and then consider that this sequence may be different when the dry substances are heated together from what they are when solutions interact at ordinary temperatures.

Stahl's conclusions formed the basis for the work of Tobern Olaf Bergman (1735–1784), who did much to systematize analysis and was one of the most fervent adherents of the theory of affinity. Bergman referred to affinity as elective attraction and believed that it was the result of the attraction between the smallest particles of matter.

It is found that all substances in nature, when left to themselves and placed at proper distances, have a natural tendency to come into contact with one another. This tendency has long been distinguished by the name of "attraction." I do not propose in this place to inquire into the cause of these phenomena, but in order that we may consider it as a determinable power, it will be useful to know the laws to which it is subject in its operations though the mode of agency be as yet unknown.[6]

Bergman held that the chemical affinity of each substance was of a definite and constant strength and that it was also selective in that it manifested itself only in a unidirectional fashion and only between determinate substances. Since he lacked the means for actual measurement, Bergman referred back to Geoffroy's tables and arranged affinities in the order of their magnitudes. He defined the order of attraction as follows:

Suppose *A* to be a substance for which other heterogeneous substances *a*, *b*, *c* have an attraction; suppose further *A* combine with *c* to saturation (this union I shall call *Ac*) should upon the addition of *b* tend to unite with it to the exclusion of *c*, *A* is then said to attract *b* more strongly than *c*, or to have a stronger

[5] Ostwald, *Lehrbuch der Allgemeinen Chemie*, 19.
[6] Bergman, *A Dissertation on Elective Attractions*, 3.

elective attraction for it; lastly let the union of *Ab* on the addition of *a* be broken, let *b* be rejected, and *a* chosen in its place, it will follow that *a* exceeds *b* in attractive power, and we shall have a series *a*, *b*, *c* in respect of efficacy.[7]

Bergman next concluded that this order of attraction is constant, but he also recognized that two sets of tables of elective attraction were needed, one for solutions at ordinary temperatures and one for the interaction between solids at high temperatures. It is interesting to note that Stahl's and Bergman's insight that affinity was indeed dependent on temperature was confirmed in 1783 by Lavoisier, who insisted that Geoffroy's tables had been produced at a constant temperature and that the tables would have looked different had the temperature been altered.

According to Bergman, elective affinity was an invariable force and of such a nature that a body that expelled another from its combination could not possibly be separated from the same by the body that it had eliminated.

> If three substances are mixed, the two whose mutual affinity is strongest combine, to the exclusion of the third. If a compound of two substances is mixed with a third substance, no reaction takes place when the affinity of the original compound for the free substance is stronger than that of either of its constituents for the free substance. If the free substance possesses an affinity for one of the constituents that is stronger than the affinity of the compound, it combines completely with that substance, displacing an equivalent amount of the other constituents.[8]

Thus, the salient conclusions derived by Bergman regarding "elective attraction" or affinity are that (1) there is a sequence in the magnitude of the elective affinities of a series of substances toward one with which they all combine; (2) this is manifested by the fact that the one possessing the greater affinity expels from the combination the one possessing the lesser affinity; (3) the order of affinity is constant under each of the two different conditions of interaction (ordinary temperatures and high temperatures) in the moist and dry way respectively, but differs under the two conditions of interaction; (4) the substance of lesser affinity is completely expelled by that of greater affinity; and (5) it is impossible to reverse such a reaction.

In reaching these general conclusions, Bergman was not proposing hypotheses but was simply reflecting his own experimental observations, and these are

---

[7] Ibid.

[8] Holmes, "From Elective Affinities to Chemical Equilibria: Berthollet's Law of Mass Action," 105–106.

precisely the conclusions that Berthollet would later attack. Berthollet "was opposed Bergman's idea that when two substances combined they would have a selective affinity for each other and would always combine in constant proportion. Berthollet maintained that affinity was a relative concept, which varied with the physical conditions accompanying a reaction."[9] He introduced the idea of the dependence of affinity on "chemical mass" or what today we would call concentration.

> The most important reactions to which chemists applied the concept of elective affinity were those between acids, bases, and salts. Believing that a salt is the direct combination of an acid and a base, they considered the reaction between a salt and an acid to be a displacement of one acid by another in combination with the base. According to the theory of elective affinities, an acid will replace another if its affinity for the base is stronger than that of the previously combined acid for the base. Similarly, a base will replace another base if its affinity for the acid was stronger ... Reactions between two salts were considered to be double-displacement reactions, resulting in an exchange of bases between the respective acids.[10]

Berthollet's Concepts of Chemical Compound and Variable Proportions

Claude-Louis Berthollet (1748–1822) was one of the most brilliant of Lavoisier's distinguished contemporaries and successors. He had been an early adherent of Lavoisier's oxygen theory of combustion, having rejected the phlogiston theory after experiments with chlorine had led him to agree with Lavoisier that all acids were necessarily oxygenated compounds. Ultimately, Berthollet advanced modern chemistry more than anyone before him, with the obvious exception of Lavoisier. Berthollet was also a staunch critic of Bergman's experimental conclusions on the nature of affinity and published his arguments in his paper *Recherches sur les lois de l'affinité* (1799), which was the starting point of the complete new system of chemistry that he later developed in *Essai de Statique Chimique*. In his 1799 paper, Berthollet "intended to distinguish the forces that acted in chemical reactions, which included not only affinity but also the forces that operated together with affinity, either in producing combinations or in bringing about decompositions."[11] He states:

[9] Fujii, "The Berthollet-Proust Controversy and Dalton's Chemical Atomic Theory 1800–1820," 177.

[10] Holmes, "From Elective Affinities to Chemical Equilibria: Berthollet's Law of Mass Action," 106.

[11] Wisniak, "Claude-Louis Berthollet," 53.

[A]ll chemical phenomena are the result of the reciprocal action that molecules exert on one another by reason of their nature; this action depends on attractive forces that tend to join them and repulsive forces that tend to separate them. . . . All combinations and compositions are acted on by two forces: by an attraction acting at a distance in which the quantity will change the effects, and a surface attraction upon which the dimensions and shape of the salts that form the combination will exert an influence. . . . This type of attraction should not be confused with general attraction [which is] the source of weight and of celestial phenomena.[12]

Berthollet's main thesis in this work was that chemical action was due partly to affinity, which he called "the principle of the intimate action of the molecules of materials"[13] and which he regarded as similar to gravitational attraction and as partially due to the masses of reactants. Here, he also expressed doubts regarding the reality of constant composition of chemical compounds, an idea that would later be refuted by Joseph Louis Proust, as we shall see presently. However, the course of his experimental work led Berthollet to further refine his theory of the dependence of chemical reactions on affinity and masses. He begins by stating that

if I can prove that a weaker degree of affinity can be compensated by an increase of quantity, it will follow that the action of any body is proportionate to the quantity of it which is necessary to produce a certain degree of saturation. This quantity, which is a measure of the capacity of saturation of different bodies, I shall call mass. . . . When a substance acts on a combination, the subject of the combination divides itself between the two others, not only in proportion to the energy of their respective affinities, but also in proportion to their quantities.[14]

Berthollet noted that, "during the extraction of saltpeter from crude niter rock by dissolution in water, the increasing concentration of saltpeter in solution made the remaining portions more difficult to dissolve, even though the water was never saturated with salt. Further quantities could be dissolved easily by using fresh water, but each successive washing yielded a smaller amount."[15] Berthollet explained this reaction by as being the result of the great quantities of sodium chloride and calcium carbonate. As the products were continuously removed, the calcium chloride was lost through seepage below ground. The

[12] Berthollet, Éssai de statique chimique.
[13] Berthollet, "Séances des écoles normales recueillies par des sténographes et revues par les professeurs."
[14] Berthollet, Recherches sur les Lois de l'Affinité.
[15] Holmes, "From Elective Affinities to Chemical Equilibria: Berthollet's Law of Mass Action," 109.

large quantities in solution of the original substances compared to those of the products produced a reaction that would not occur by affinities alone. When a salt solution filtered slowly through the pores of the limestone, the relatively weak affinities between these two substances were enhanced by the combined effects of the temperature and the enormous mass of limestone. This led to decomposition of the salt, assuming a constant production of sodium carbonate and calcium chloride through double decomposition.[16]   .

These results and conclusions have led to renewed interest in Berthollet for his earlier suggestion that physical conditions such as temperature, relative concentration, and quantities of reactants affected the nature and direction of affinities in a chemical reaction. Thus, Berthollet proposed that

> affinities do not act as absolute forces, so that one substance can be directly expelled by another from its compounds; but [holds] that in all combinations and decompositions which are the result of affinity, the substance on which two others act with opposing forces, always divides itself between these two substances, and that the ratio of the division does not only depend on the strength of the affinities, but also on the quantities of the active substances present; since to produce an equal division, the greater mass of one substance can compensate for what is lacking in affinity.[17]

In turn, Pierre-Simon Laplace considered that his Newtonian theory of chemical affinity received support from Berthollet's demonstration that the effective affinity of a reactant depended on the quantity or mass present. Additionally, Berthollet showed that if two acids were present with a base, it was not simply the stronger acid that combined with the base, but instead a partition took place in which each of the acids reacted with the base in certain proportions, which Berthollet attributed to their respective affinities. Laplace argued that this confirmed the action of short-range forces and that contact alone would not give this result.

Berthollet affirmed that the "force of affinity depended on the relative distance between the particles composing a substance: The more closely packed these were, the stronger their reciprocal affinity. Consequently, it was more difficult for substances to combine in the solid state than in the liquid [state]. In the former case, the particles of a given substance had to overcome their affinity for each other before combining with the particles of another substance."[18] In the latter

[16] Holmes, "From Elective Affinities to Chemical Equilibria: Berthollet's Law of Mass Action."
[17] Berthollet, *Recherches sur les Lois de l'Affinité*, 113.
[18] Kapoor, "Berthollet, Proust, and Proportions," 61.

case, the particles were further removed from each other and were more susceptible to attracting the particles of a foreign body.[19]

Essentially, Berthollet's starting point was the idea, now known to be false, that
Newton's law of universal attraction also applied to what today would be called
molecules. He claimed that all the forces that produce chemical phenomena are
derived by the mutual attraction of bodies, which had been given the name "affinity" in order to distinguish it from the attraction that governs the heavenly
bodies. We understand today that this model does not correctly describe reality,
but Berthollet could have no conception of the microscopic scale of the phenomena with which he was dealing.

According to Berthollet, three types of affinities could be considered as different states of the same power: simple attraction, elective attraction, and complex attraction. *Simple attraction* was the type of affinity exerted between two
substances whose forces interacted with each other, although the substances are
composed of different principles. *Elective attraction* was the property of a body
chasing away another body from a combination in order to take its place, so it
was involved in the competition between two substances for a third substance.
Complex attractions took place when more than two such bodies were present.

Berthollet proceeded to supply what he considered to be the experimental
basis for his claims. He chose many of his examples from the class of reactions
in which alkali and alkaline earths interact with acids. His experiments had
a bearing on two distinct points. First, they proved that the result of a chemical reaction depends not only on affinity but also on the relative quantities of
interacting substances. Second, by investigating the conditions that modify the
application of this principle, he showed that the mass of a substance undergoing
chemical reaction becomes relevant only if each of its particles can act and react
with each particle of the other substances participating in the reaction. Hence,
any substance that, owing to its insolubility or volatility, is removed from a solution in the form of a precipitate or gas cannot be considered to contribute to
the "active mass" involved in the reaction. Therefore, the solubility and volatility
of the substances involved in a reaction have a profound influence on the final
result of that reaction. Without realizing it, Berthollet seems to be introducing
the concept of equilibrium reaction, although he does not believe that a reaction
occurs in both directions, from reagents to products and vice versa.

Berthollet proceeded to illustrate these conclusions with many examples
displaying the influence of solubility on the final products of chemical reactions.
Among these examples was the formation of barium sulphate. He claimed that,
in comparing the affinity of baryta (barium hydroxide) toward sulfuric acid
with that of potash (soda) toward this same acid, the influence of solubility is

---

[19] Sadon-Goupil, *Révue de l' Éssai de statique chimique.*

enormous. Owing to its almost complete insolubility, the barium sulfate is always removed from the solution as soon as it is formed. Consequently, in the joint competition between the potash and baryta for the sulfuric acid, the alkali obtains only an imperceptibly small share of the acid, and, if one were to make this the basis for evaluating the relative affinities, one would reach erroneous conclusions.

The discussion of this case and similar ones led Berthollet to recognize that the insolubility or lesser solubility of a substance was equivalent to the reduction of its active mass. If, among the possible combinations between the constituents participating in a reaction, there is one that is insoluble or less soluble than the others, that constituent will be formed at the expense of all the others. Thus, the insoluble carbonates of calcium, barium, and strontium are precipitated when soluble salts of these elements are mixed with soluble carbonates. From aqueous solutions containing several salts, the salts will be precipitated according to the order of solubilities, rather than to the actual amounts of constituent salts present in the solution. However, since the solubility of salts alters with changing temperatures, the temperature at which the reaction is taking place must always be considered. Berthollet concluded, as Bergman had also understood, that affinities could not be subjected to any measuring procedure and he, thus, set aside the tables of affinity.

These ideas led Berthollet to conclude that compounds do not unite in definite proportions but that, instead, they combine in variable proportions according to physical circumstances under which they are combining. This idea was due to his belief that other forces were involved in chemical reactions besides affinity, temperature, and solubility. These included the mass of the reactants, the cohesion, and the elasticity in different states of matter. He suggested that chemical reactions seek a static equilibrium. So, for example, since he believed that the molecules in a gaseous reaction were separated by caloric, such reactions must be carried out at certain temperatures in order to succeed.

These ideas led Berthollet to challenge the notion of the constant composition of compounds by pointing to the considerable body of evidence suggesting that compounds had variable composition. For example, when some metals are heated, they form oxides that appear to have a regularly changing percentage of oxygen found in the oxide. Perhaps in certain cases a given mass of alkali may always seem to neutralize the same quantity of acid, just as the quantity of sea salt dissolved by a given mass of water depends only slightly on the temperature. But, according to Berthollet, these are special cases and cannot be taken as distinctive of a general law. The proportions in which the elements come together in a combination are not fixed and usually depend on the conditions under which the combination occurs.

Although Berthollet emphasized that all substances combine in indefinite proportions, special conditions prevent substances from combining in variable proportions. These conditions include condensation, volatilization, and precipitation. Nevertheless, his claims supporting variable proportions were not borne out by experiment, as was emphasized by the chemist Joseph Louis Proust. Proust disagreed vehemently with Berthollet on this issue and maintained that chemical combinations follow definite proportions.

Proust's Classification of Substances and the Law of Definite Proportions

Joseph Louis Proust (1754–1826) made notable contributions to many areas of chemistry, ranging from the method for fabricating powders to the analysis of pharmaceuticals for treating yellow fever and the manufacture of soaps and grape sugars. However, he is most influential in the history of chemistry for establishing the Law of Definite Proportions, which stipulated that each chemical combination has a composition that is fixed, specific, and rigorously independent of the conditions under which the combination takes place. This idea revolutionized the notion of mixt because it negated what Newton and his followers had taken for granted regarding chemical saturation. However, the development of this law also marks one of the definitive turning points in the recognition of chemistry as a genuine and systematic science. It is important to understand that skepticism over the scientific status of chemistry was not restricted to Cartesians, who questioned the heuristic value of experimental work and who sought to reduce all science to a priori principles derivable from pure reason. The scientific status of chemistry was also denied by philosophers such as Immanuel Kant, who claimed that

[w]hat can be called a *proper* science is only that whose certainty is apodictic; cognition that can contain mere empirical certainty is only *knowledge* improperly so-called. Any whole of cognition that is systematic can, for this reason, already be called *science*, and, if the connection of cognition in this system is an interconnection of grounds and consequences, even *rational* science. If, however, the grounds or principles themselves are still in the end merely empirical, as in chemistry for example, and the laws from which the given facts are explained through reason are mere laws of experience, then they carry with them no consciousness of their *necessity* (they are not apodictically certain), and thus the whole of cognition does not deserve the name of a science in the

strict sense; chemistry should therefore be called a systematic art rather than a science.[20]

This last point recalls the modern idea that the presence of a body of necessary laws is required to establish a true and systematic science. Thus, for chemistry to earn the status of a science proper, a true Galilean science, it must be grounded in necessary laws that can be confirmed by experimental observations. In this regard, Proust understood natural laws as being primitive, immediate, universal, and invariable. He thus busied himself with finding such laws that govern chemical processes and reactions.

One of the central problems that concerned Proust was how to distinguish between compounds and mixtures in order to distinguish elementary analysis from immediate analysis, and it is to this end that he developed the Law of Definite Proportions. Up to and including the time of Lavoisier, the term *mixt* had included both mixtures and compounds. Furthermore, Lavoisier treated the decomposition of air and water in parallel fashion, suggesting that these processes were analogous. Nevertheless, Proust considered it crucial to distinguish the elementary analysis of pure compounds such as water from the immediate analysis of a mixt, such as air, into its pure constituents.

Proust adopted an experimental criterion to distinguish mixts from compounds, which is the invariability of their quantitative elementary composition. In 1799, he provided strong evidence against Berthollet's thesis that solutions, alloys, and gases are all chemical compounds with indefinite proportions. He also demonstrated that, if natural copper carbonate is dissolved in an acid and then precipitated by an alkaline carbonate, the amount of copper carbonate that is obtained is exactly equal to the natural carbonate used at the start of the experiment. The transformation produces neither a gain nor a loss of carbonic acid or copper oxide. This provides strong evidence that the composition of copper carbonate prepared in a laboratory is identical to that found in nature, although the chemist's process of preparation is completely different from nature's process.

This and other experimental work led Proust to publish numerous papers between 1802 and 1806 in the *Journal de Physique, de Chimie, et d'Histoire Naturelle*, in which he attempted to demolish Berthollet's ideas regarding proportions in compounds. Of all the articles published by Proust, the ones from 1804 stand out for their most overwhelming response to Berthollet's imprecise observations. In particular, these articles call attention to the lack of experimental observations that would illustrate Berthollet's conclusions, and, in these articles, Proust defends his basic postulate that one should never elaborate theories that are contrary to the facts.

[20] Kant, *Metaphysical Foundations of Natural Science*, 4.

On the basis of these discoveries, "Proust did not hesitate to affirm that all chemical combinations are characterized by an absolutely fixed, specific combination independent of the conditions in which the combination is formed. All the cases of combination with variable composition that have been through to occur are, in fact, impure combinations containing an excess of one of the components, or a mixture of two distinct combinations of the same elements."[21] Proust states that "these eternally invariable proportions, these constant attributes that characterize the true compounds of art, or those of nature, in a word the *pondus naturae* so well understood by Stahl; all this, I say, is no more in the power of the chemist than the law of election which governs all combinations."[22] Proust concluded from his experimental observations that the distinction between solutions and compounds was clear and that "nature has imposed upon itself certain laws of proportion according to which it forms these unions that we conveniently call combinations.[23]

In formulating the Law of Definite Proportions, Proust had an invaluable impact on John Dalton's work on atomic weights and on his subsequent development of chemical atomism. The Law of Definite Proportions posited that each chemical compound was constituted by a fixed and constant proportion of components. Although Berthollet vigorously opposed this law, which led to a long, though polite, dispute between the two chemists, it eventually succeeded over Berthollet's conjectures and influenced the development of Dalton's Law of Multiple Proportions and of Joseph Louis Gay-Lussac's Law of Volumes, through which the natural production of compound bodies found its limits in the laws of nature themselves. From that point forward, two categories of mixts were distinguished from one another: the chemical combinations and the physical mixtures. This distinction was due to the fact that the Law of Definite Proportions was not applicable to physical mixtures but was held rigorously in chemical combinations. It was thus to become the criterion that permitted mixtures to be distinguished from chemical combinations.

Although 18th-century chemistry's excessive emphasis on the decisive role of production methods had jeopardized the universal character of science, the work of Proust, Dalton, and Gay-Lussac finally established the uniformity of nature, even within the artificial conditions of the laboratory. Their work, however, also established that nature is discontinuous because the laws of weight and proportion are not only universal but also "discrete." As a result, the doctrine of affinity was soon forgotten, and chemistry could finally begin making great progress within a relatively short period of time.

---

[21] Duhem, *"Mixture and Chemical Combination" and Related Essays*, 41.
[22] Proust, "Recherches sur le cuivre," 26.
[23] Proust, "Sur les mines de cobalt, nickel et autres."

6

John Dalton and Chemical Atomism

The Problem of Reconciling Elementarity with Atomicity

Even after the Renaissance rediscovery of the classical atomistic conception of matter and its reappraisal by early modern natural philosophers, the relevance of atomism for chemistry remained unclear for two reasons, which were interposed once the applicability of atomism to chemistry was considered. The first of these reasons was practical, and the second reason was conceptual. From the practical point of view, mechanistic atomism did not regard atoms as having properties that were experimentally identifiable, measurable, or quantifiable. From the conceptual point of view, classical atomism clashed with both the theory and practice of chemistry because it represented atoms as having uniform properties. The fundamental particles of Democritus, Gassendi, Boyle, and even Newton were all of one type since they had no properties other than shape, size, and motion or rest, though for Newton they also displayed forces of attraction and repulsion.

Early modern chemists well knew, however, that homogeneous chemical substances were qualitatively distinct from each other, and it was not clear how such observable qualitative differences could be explained by the uniform microscopic properties of atoms. To say, as Boyle had done, that distinct observable properties were due to a difference in the stable microstructural arrangement of corpuscles would have required explaining how such a microstructure could maintain its stability in the absence of bonding forces. This was, in fact, a major problem associated with Boyle's account of microstructure or texture, as he called it. Another area of concern for chemists was how to understand the possible relationship between the fundamental parts of matter and what had been identified, at least provisionally, as elementary substances.

In addition to these issues (and as can be gleaned from the discussions of atoms and elements in the previous chapters), it was also not clear throughout most of the history of natural philosophy and early modern science how the concept of atom could possibly be related to the concept of element, since these notions implied completely different ways of conceptualizing fundamental entities or substances. The concept of fundamentality can be understood in either of two ways. *Fundamentality* can refer to the more general metaphysical concern with the ultimate nature of reality, and this is the way in which the atomists understood this concept. However, fundamentality can also refer to a narrower

From the Atom to Living Systems. Marina Paola Banchetti-Robino and Giovanni Villani, Oxford University Press.
© Oxford University Press 2023. DOI: 10.1093/oso/9780197598900.003.0007

ontological concern with how observable entities and chemical substances can be hierarchically organized in terms of ontological dependence. This second way of understanding 'fundamentality was applied to elements throughout most of the history of Western philosophy and early modern science. However, to the extent that these two senses of fundamentality are not necessarily co-extensive, the related concepts of atoms as metaphysically fundamental and of elements as ontologically fundamental were quite difficult to reconcile.

To recapitulate briefly what was discussed in detail in the previous chapters, from the classical to the early modern period, elements were generally regarded as the principles upon which the properties of existing things depended. As such, elements were not considered irreducible or metaphysically fundamental. They were, in fact, regarded as being (at least in principle) observable substances with determinable properties. The complex properties of material bodies were considered to be ontologically dependent on the simple properties of the elements that compose them.

On the other hand, in both its classical and its early modern iteration, atomism was a metaphysical thesis purporting to establish claims about the ultimate nature of material reality. Advocates of this theory regarded atoms as being irreducible and metaphysically fundamental. Since such entities were considered to be unobservable even in principle and since their properties were not empirically determinable, the belief in the existence of atoms was founded more on speculation than on sound empirical evidence. Thus, rather than being an empirical notion, the concept of atom was the idea of ultimate substances that grounded all material beings. Although Lavoisier had shunned atomism and metaphysical speculations about the nature of fundamental matter in order to focus on empirically identifiable elementary substances, the desire to explain what it was that made such substances elementary would linger in the minds of chemists, and it begged for some heuristic account.

These questions were precisely what motivated John Dalton (1766–1844) to seek a reconciliation of the atomistic conception of matter with Lavoisier's operational notion of element, while at the same time providing quantitative information about the properties of atoms that would explain the qualitative differences between the homogeneous substances that were identified as elements. Dalton proposed *measuring* the weights of atoms indirectly, thereby both confirm the atomic hypothesis and transform it into an atomic theory. When Dalton discovered how to differentiate atoms on the basis of weights, the atomic theory would finally have a real effect on the development of modern chemistry.

Although the operational conception of chemical elements advanced by Lavoisier had deliberately separated the macroscopic plane from the experimentally inaccessible microscopic plane, Dalton succeeded in reconciling the two planes in a manner that was both quantitative and empirical, that is, supported by experimental evidence. The very co-presence of the microscopic and macro-scopic planes of atoms, molecules, elements, and compounds, which was made possible by Dalton's chemical atomism, would eventually constitute the defining character of chemical explanation and would ultimately lead to the discovery of valence and to the 1869 publication of Mendeleev's periodic table of the elements in *Zeitschrift für Chemie*.

The Weakness of Atomism for Experimental Science

Before discussing the development and advantages of Dalton's chemical atomism, it is important to examine the problems that motivated many 17th- and 18th-century natural philosophers to disfavor corpuscularian and atomistic theories of matter. Although early modern particulate matter theorists clearly believed that atoms and corpuscles have empirical reality, actual empirical support for atomism and corpuscularianism was lacking in the 17th and 18th centuries. As Christoph Meinel aptly points out, "in atomism . . . there was no experimental proof possible, although most corpuscular theories of the seventeenth-century explicitly claimed to be based upon experience."[1]

The (mostly unconvincing) arguments presented in support of atomism fell into four general categories. The first group of arguments, which can also be found in the writings of ancient Epicurean atomists, involved analogically extrapolating the existence of atoms and their minute dimensions from mac-roscopic phenomena. Looking back to Sennert's 1636 experiment in which he "described a distillation in which a stream of alcohol vapor passed through a sheet of paper, the density of which was supposed to give an idea of how small the atoms really were . . . it is clear that there was no quantitative methodology behind these indications. Sennert used the language of the laboratory in a merely figurative and persuasive manner, appealing to the imagination of the reader."[2]

The second type of argument in favor of atomism was based on microscopic observations. Since the invention of the compound microscope in the 1590s, early modern scientists had been fascinated with "the worlds to be found in a drop of water,"[3] although it wasn't until the publication of Giambattista Odierna's *L'occhio della mosca* (*The Fly's Eye*) in 1644 that the first detailed account of mi-croscopic observations became available to the larger scientific community. It was also not until the 1660s that the microscope became widely used for scien-tific research. Gassendi was one of the first natural philosophers to realize the

---

[1] Meinel, "Early Seventeenth-Century Atomism: Theory, Epistemology and the Insufficiency of Experiment," 68.
[2] Ibid., 78–79.
[3] Ibid., 80.

scientific potential of this instrument, although this potential "was presented as an empirical fact by Henry Power a few years later."[4] However, as Power's own attempts demonstrated, the potential of the 17th-century microscope as a research tool was more likely to be fulfilled in areas such as biology than with regard to the theory of matter. The best one could do was to extrapolate the existence and dimensions of atoms from what was observed via the microscope. "Since [atoms] are so small that they could be inferred only rationally, methodological difficulties . . . arose if one attempted to model the real after the visible."[5]

The third type of argument in favor of atomism and corpuscularianism concerned "transport phenomena in which material substances appear or disappear invisibly. . . . [For example] [t]he drying of bread or the slow evaporation of liquids are material processes, although the flux of material cannot be observed. In all these cases, quantitative change can be recorded, and from this the existence of invisible parts of matter could be inferred."[6] However, although this type of naively inductive argument rendered plausible the existence of atoms or corpuscles, it could neither prove their existence nor establish the material identity of these particles.

The fourth type of argument, based on physical experiments involving changes such as condensation and rarefaction, attempted to provide more convincing empirical evidence for the existence and behavior of atoms. These experiments remained inconclusive, however, because their results were open to interpretation regarding the existence or nonexistence of the vacuum. The ancient atomists had posited a vacuum to explain the mobility of hard and impenetrable atoms and, thus, the possibility of change. This vacuum could be either a continuous void or an interparticulate void. The existence of the vacuum was extremely controversial in the 17th century. Even after Evangelista Torricelli (1608–1647) produced the first vacuum in 1643 in a column of mercury and under controlled experimental conditions, the implications of this experiment for corpuscularian matter theories were unclear.

No clear consensus could be reached on this issue, with some philosophers arguing for the continuous universal vacuum, others arguing for interparticulate vacua, and still others arguing against the existence of any vacuum at all and favoring a substance, such as the ether, that both filled the spaces between atoms and held the particles of bodies together.[7] However, as Meinel points out, "the reintroduction of an active spirit or ether into atomism, aimed at explaining how the atoms interact and how their actions are transmitted, undermined the theoretical consistency of the mechanical corpuscularianism. . . . Plenist corpuscular

---

[4] Ibid.
[5] Ibid.
[6] Ibid., 81.
[7] Ibid., 82.

theories such as [Sebastien] Basso's and Descartes' exemplify that it was entirely
acceptable to assume corpuscles without admitting the void."[8] Still, the physical experiments themselves could not settle this question, since the very results
of these experiments could be interpreted to support any of the competing
hypotheses regarding the vacuum. It is no wonder, then, that Cartesian mechanical philosophers were suspicious of the usefulness of experiment for confirming
the mechanical hypothesis and ultimately sought to replace the uncertainty of
laboratory practice with the rigorously mathematized discourse of physics.[9]

By the mid-18th century, the only way to resolve the issue of what is to be
regarded as fundamental from the perspective of an experimental science
like chemistry was to follow Lavoisier in denying the value of metaphysical
speculations and a priori conjectures and to focus only on what substances
could be identified as fundamental in the laboratory. This was the case even if
these conclusions were only provisional and open to revision as the methods of
chemical analysis evolved and identified even more fundamental substances.
Furthermore, the requirement for quantification that was rightfully regarded as
the hallmark of rigorous empirical science was to be achieved by careful measurement and weighing. Yet, as already suggested, the question of what contributed
to the diversity of elementary substances remained open. The answer would be
postponed until John Dalton reconceptualized the atom itself.

John Dalton

Dalton's *New System of Chemical Philosophy* (1808) affirms his commitment to
Lavoisier's operational and analytical definition of element, which had become
conventional for chemists by the beginning of the 19th century.[10] In a manner
similar to Lavoisier, Dalton states that "by elementary principles, or simple
bodies, we mean such as have not been decomposed, but are found to enter
into combination with other bodies. We do not know that any one of the bodies
denominated elementary, is absolutely indecomposable, but it ought to be called
simple, till it can be analyzed."[11] However, in spite of agreeing with Lavoisier regarding elementary principles, Dalton's meteorological studies on the properties
and behavior of gases prompted him to ask a number of important questions
that ultimately led to his inquiries about the compositional nature not only of
gases but of all elementary substances. He asked: "Why do different gases have
different solubilities in water?," "Why are light and elementary gases such as

---

[8] Ibid., 83.
[9] Bernard, "Le cartésianisme de Boyle," 155.
[10] Boas Hall, *John Dalton and the Progress of Science*, 21.
[11] Dalton, *A New System of Chemical Philosophy*, 222.

hydrogen and oxygen least soluble, while compound gases such as carbon dioxide are very soluble?," and "Is solubility proportional to density and complexity?" These questions fueled Dalton's desire to ascertain the compositional nature of gases and elementary substances but to do so in a way that would meet Lavoisier's strictly empirical and quantitative requirements. Actually, Dalton went beyond Lavoisier by fusing the theoretical aspect of atoms with the empirical nature of elementary substances in a way that, for the first time in the history of chemistry, brought the world of atoms into the world of the laboratory.

Dalton ultimately decided on the assumption, which he believed was supported by observations, that gases and chemical elements are composed of "ultimate particles" or atoms that are qualitatively differentiated. He conceptualized atoms as dense spherical particles, each of which is surrounded by a subtle fluid or caloric that prevents the particles from being drawn into actual contact with one another. He claimed that this is proven by the observation that the bulk of a body may be diminished by abstracting some of its heat.[12] It is noteworthy that this is the material version of Lavoisier's and Laplace's dual material or kinetic theory of heat. Dalton's theory of change of state did not take into account a latent heat, which was formulated by Joseph Black around 1761, following evidence from William Cullen in 1756 on the cooling of solids when liquids evaporated from their surface. It also does not take into account the evidence from sources, including George Martine in 1740, that specific heat did not depend on the weight of the surface.

Dalton regarded chemical reactions as involving the shuffling and reshuffling of atoms into new clusters and, influenced by Newton's idea of forces of attraction, he wrote that

> observations have tacitly led me to the conclusion which seems universally adopted, that all bodies of sensible magnitude, whether liquid or solid, are constituted of a vast number of extremely small particles, or atoms of matter bound together by a force of attraction, which is more or less powerful according to circumstances, and which ... endeavours to prevent their separation, and is very properly called ... *affinity.* ... Besides the force of attraction ... we find another which comes under our cognizance, namely, a force of repulsion.[13]

Although the view expressed here may seem like a return to the speculative atomism that Lavoisier had rejected, Dalton buttresses his chemical atomism by postulating that there are as many distinct atoms as there are distinct chemical elements. He assumes that atoms of the same element are similar in shape

---

[12] Ibid., 144.
[13] Ibid., 141–143.

and mass but differ from atoms of other elements. He also postulates that the properties of a compound substance are determined by the properties of the elements of which it is composed and that the properties of each element are determined by the properties of its distinctive constituent atoms. He thus reconceptualizes the notion of atom and of element by subsuming both under the concept of fundamentality, which is understood as ontological dependence.

Dalton regards the antagonism between the forces of attraction and repulsion affecting atoms, as well as the quantities of heat surrounding each atom, as accounting for the differences between gaseous, liquid, and solid bodies.[14] He considers that gases consist of corpuscles that repel each other when exposed to heat and that one must differentiate these corpuscles not only by size and shape, as claimed by classical atomists, but also by weight.[15] Regarding the solubility of gases, Dalton concludes that "the circumstance depends upon the weight and number of the ultimate particles of the several gases: Those whose particles are lightest and single being least absorbable, and other more according to their increase in weight and complexity."[16] Although he was a 19th-century scientist and did not operate within the theoretical framework of contemporary chemistry, Dalton seems here to already be intuiting that the problem at hand is not merely quantitative but also qualitative.

One can see how this line of reasoning would then lead Dalton to pursue an inquiry into the relative weights of chemical atoms.[17] In seeking to conduct such an inquiry, Dalton eliminates the intangibility implied by metaphysical atomism and has fixed a determinable property to chemical atoms. Dalton believes that the relative weights of atoms can be determined by measuring the relative weights of elements that are isolated during the analysis of compounds. Therefore, availing himself of the highly precise data provided by Lavoisier's research,[18] Dalton proposes to use analysis to isolate the component elements of compound substances and to measure their relative weights. He explicitly states that "it is one great object of this work, to shew the important advantage of ascertaining *the relative weights of the ultimate particles both of simple and compound bodies, the number of simple elementary particles which constitute one compound particle, and the number of less compound particles which enter into the formation of one or more compound particles.*"[19]

In order to move from measurement of the relative weights of elements to the relative weights of the constituent atoms, Dalton had to make a number of other

---

[14] Ibid., 144.

[15] Bensaude-Vincent and Stengers, *Histoire de la Chimie*, 113.

[16] Dalton, "On the Absorption of Gases by Water and Other Liquids," 217–287.

[17] Leicester, *The Historical Background of Chemistry*, 155.

[18] Newman, *Atoms and Alchemy: Chymistry and the Experimental Origins of the Scientific Revolution*, 221.

[19] Dalton, *A New System of Chemical Philosophy*, 213 (emphasis in the original).

assumptions, this time about chemical composition and chemical reactions. One of these assumptions is the Law of Conservation of Mass, which had been previously recognized by Jan Baptista van Helmont and which was considered "approximately true" or "empirically true" by the 19th century. Dalton concludes that atoms can neither be created nor destroyed in chemical reactions, and he states that "chemical analysis and synthesis go no farther than to the separation of particles one from another and to their reunion. No new creation or destruction of matter is within reach of chemistry."[20]

Dalton also assumed the Rule of Greatest Simplicity, according to which chemical formulas always take the simplest form that is compatible with the empirical data. Therefore, Dalton postulates that "if there are two bodies, A and B, which are disposed to combine, the following is the order in which the combinations may take place, beginning with the most simple [combination]: namely, binary [AB], ternary [A^2B or AB^2], quaternary [A^3B or AB^3], etc."[21] Dalton has no way to explain how or why such combinations occur, and his error here is clearly that he assumes a "combinatory logic" between inert material entities.

Dalton's theory also places Joseph Proust's Law, which is today known as the Law of Definite Proportions, on a firm theoretical ground. The Law of Definite Proportions dictates that a chemical compound always contains exactly the same proportion of elements by mass. Dalton grounds this law by stipulating that compounds are made of combinations of different types of atoms in fixed proportions, regardless of the source or method of preparation.

In addition to adopting the Law of Definite Proportions, he also formulated the Law of Multiple Proportions such that, when two elements combine to form two or more compounds, the ratios of the masses of one element that combine with the fixed mass of the other are simple whole numbers. As an example, we can cite carbon monoxide (CO) and carbon dioxide (CO_2).[22]

Armed with these postulates and with empirical data provided by chemical analysis and measurement using the precision balance, Dalton proceeded to calculate the relative weights of chemical atoms from the experimental data about compound substances, using hydrogen as the fixed unit so that "the atomic weight of each element [is] the gravimetric proportion that combine[s] with a gram of hydrogen to form the most stable combination."[23] After having collected enough data on relative weights, Dalton moved on to create his famous tables of elements. These tables grew larger and more complex over the course of 24 years as Dalton identified more and more elements based on relative

[20] Newman, *Atoms and Alchemy: Chymistry and the Experimental Origins of the Scientific Revolution*, 213.
[21] Ibid.
[22] Ibid.
[23] Bensaude-Vincent and Stengers, *Histoire de la chimie*, 114.

weights and atomic masses. By the time the second volume of the *New System of Chemical Philosophy* was published in 1827, the number of elements had grown to thirty-six.

Thus, although basic substances (or atoms) do not necessarily overlap with simple substances (or elements), Dalton succeeded in bringing the notion of compositional fundamentality together with that of explanatory fundamentality in the context of his chemical atomism. These two notions are reconciled by showing that the chemical properties of elements are both mereologically and causally dependent on the empirically determinable properties of their constituents' chemical atoms. For the first time in the history of chemistry, atoms were associated with a numeric value that could be obtained using practical experiments. The philosophical importance of this advance is that it contributed to definitively changing atomism from a merely philosophical hypothesis to a properly scientific theory.

Some Problems with Dalton's Chemical Atomism

As Alan Rocke explains and as has been discussed in this chapter, there is a clear distinction between chemical atomism, which "forms the conceptual basis for assigning relative weights to elements and assigning molecular formulas to compounds,"[24] and the physical atomism expounded by both Democritean and early modern atomists. Dalton himself embraced both theories, that is, the theory of chemical atoms as compositionally and mereologically fundamentals and as explanatorily and causally fundamental. The clear relation between chemical atomism and physical atomism in his work generated much controversy among 19th-century chemists, for whom physical atomism continued to be "far from universally accepted, since it made what seemed to many to be rather doubtful statements about the intimate mechanical nature of substances."[25] For these reasons, many of Dalton's contemporaries continued to sympathize with Lavoisier's view that all atomic theories were ultimately bound to be metaphysical.[26] "Few scientists distinguished between the two theories, and as a consequence every attack on physical atomism impugned, by association, the scientific worth of chemical atomism."[27]

Although several 19th-century chemists adopted a more pragmatic stance and chose to adopt chemical atomism as a convenient instrumental device without making any realist ontological commitments to the physical reality of atoms,

[24] Rocke, *Chemical Atomism in the Nineteenth Century: From Dalton to Cannizzaro*, 10.
[25] Rocke, *Chemical Atomism in the Nineteenth Century: From Dalton to Cannizzaro*, 10.
[26] Knight, "John Dalton (1766–1844)."
[27] Rocke, *Chemical Atomism in the Nineteenth Century: From Dalton to Cannizzaro*, 10.

other chemists rejected atomism altogether, both in its physical and chemical iterations. For example, the great physical chemist Wilhelm Ostwald held out against atoms well into the 20th century, believing that the facts of chemistry could be better accounted for thermodynamically and, thus, at the macroscopic level.[28] Because the field of statistical mechanics that linked the microscopic atomic level to the macroscopic phenomena had already been developed during Ostwald's time, it is clear that his choice of in favor of thermodynamics was linked to his positivistic philosophical perspective rather than to strictly scientific and empirical considerations.

We also now know that Dalton's rigid and arbitrary assumptions led him to reach some erroneous conclusions about composition and weights. Proust's Law of Definite Proportions, for example, was not without its detractors. One of these detractors was Berthollet, who disagreed with Proust and who believed that there is no set ratio for the elements in compounds, so that the ratio of elements that make up a compound depends on what one begins with in one's experiment. He believed that his view was supported by his experimental work, since a few compounds did in fact confirm his claim, due to the way in which the atoms in those compounds formed crystals. However, in many of his experiments, the samples Berthollet used were contaminated and thus did not reflect the correct ratios in the compounds he was using. Nevertheless, the fact that Berthollet was incorrect in this regard does not indicate that Proust's Law of Definite Proportions was accurate.

Dalton's conclusions regarding the formula for copper carbonate are mistaken because Proust's proportions focus on mass, while our own Law of Definite Proportions focuses on the number of atoms. Later chemical discoveries established that there are violations of Dalton's postulates. For instance, ferrous oxide, whose actual formula is nonstoichiometric ($Fe_{0.95}O$ so $Fe_{1-x}O$) rather than the more "ideal" stoichiometric formula (FeO), violates the Law of Definite Proportions, which requires a ratio of whole integers.

In addition, Dalton conceptualized the ontological dependence of properties on atomic mass and composition in a way that is too simplistic. Because he assumed that chemical properties depended only on combinations of atoms and because he was unaware of the dependence of chemical properties on the structural arrangements of atoms and on the specific interactions between them, Dalton was unable to predict and would have been unable to explain many types of chemical compounds that were later discovered. For example, his chemical atomic theory could not have predicted or explained the existence of isomers, substances that have the same numeric composition but different structures and therefore different properties. Some examples are constitutional isomers such as

[28] Ibid., 76.

butane and isobutane, which have entirely different melting and boiling points (since butane [C^4H^{10}] melts at $-138.4°C$ and boils at $-0.5°C$, and isobutane [C^4H^{10}] melts at $-159.42°C$ and boils at $-11.7°C$), nor could it have predicted the existence of stereoisomers, enantiomers, and diastereomers. Dalton could not have predicted the discovery of isotopes, such as chlorine 35 and chlorine 37, which are the same element but have different masses and densities.

In spite of these "errors," Dalton's atomic theory qualifies as a progressive paradigm in chemistry since it introduced a number of new and important concepts. It also rendered intelligible the hundreds of quantitative analyses of substances recorded in the chemical literature and provided a model for the long-standing assumption that compounds were formed from the combination of constant amounts of their constituents. His Law of Multiple Proportions explained the discontinuity in the proportions of elements in compounds. Very importantly, Dalton's chemical atomism suggested that the arrangement of atoms in a compound could be represented schematically in a way that indicated the combination of atoms in the compound. It provided a precise quantitative basis that vastly improved upon older and much vaguer ideas about atoms. Dalton's labeling and differentiating of atoms with quantitative measurements obtainable via laboratory practices clearly rendered atomism an experimental and scientific theory, rather than merely metaphysical speculation as it had been in the past.

The relation between Dalton's chemical atoms and his elements is analogous to the relation between Paneth's basic substances (*Grundstoffe*) and his simple substances (*einfacher Stoffe*), in the sense that Dalton's chemical atoms are the nonobservable constituents of substance, whereas elements are the form in which the properties of chemical atoms are manifested empirically. However, this is where any possible analogy ends since, for Paneth, basic substances are "noumenal" in the Kantian sense and thus do not have any empirically determinable properties.[29] For Dalton, on the other hand, chemical atoms do have the empirically determinable and measurable property of weight, and it is this reconceptualization of "atom" that successfully extracts this ancient idea from the realm of purely speculative metaphysics and places it squarely within the realm of concrete experimental science. It is in this regard that Dalton's chemical atomism is to be considered as scientifically progressive since, in spite of its erroneous assumptions, it represents the first major attempt to reconcile the empirical and quantitative elements of modern chemistry with the theory of discrete particles that compose material bodies. In this regard, then, it constitutes an important step in our understanding of matter and, thus, successfully contributes to the growth and advancement of modern chemistry.

---

[29] Ruthenberg, "Paneth, Kant, and the Philosophy of Chemistry," 83.

In spite of the advancements Dalton made in introducing the notion of chemical atom, the qualitative problem remained unsolved. This is because, if qualitatively distinct atoms (such as the hydrogen atom and the iron atom) have different weights, Dalton did not clarify the reason for which each of these atoms has its specific distinctive weight. In the next chapter, we shall see that, when William Prout (1785–1850) looked to the hydrogen atom as an elementary substance to solve this problem, he demonstrated that to explain the qualitative difference between one atom and the other, we must either ignore that difference or enter within the atom in order to find its internal structure.

7

Valence, Chemical Bonds, and the Theory of Elements

William Prout's Hypothesis Regarding Atomic Weights

After Lavoisier had compiled his list of the then known elements, other elements were added in the years following his death, leading to a large list of fundamental building blocks of matter. This large number seemed excessive to the physician William Prout (1785–1850), who eventually used Avogadro's method of comparing the relative densities and weights of gases and proposed that all atoms appeared to have weights that were exact multiples of the weight of hydrogen, a hypothesis that came to be known as Prout's hypothesis. "Recent work has shown that he was committed . . . to the idea [of] unitary theories of matter, and the complexity of the elements as early as 1810, when he was still a medical student at Edinburgh. Apparently he was much impressed by Dalton's atomic theory, by Gay-Lussac's law of combining volumes, and by the unitary ideas of Humphry Davy's *Elements of Chemical Philosophy* (1812)."[1]

In 1815 and 1816, Prout published two anonymous papers that discussed the relationship between the specific gravities of bodies in the gaseous state and the weights of their atoms.[2] In the introduction to the 1815 paper, Prout specified that all of his observations and conclusions were based on Joseph Louis Gay-Lussac's doctrine of volumes. He then proceeded to calculate the specific gravities of the four elementary gases (oxygen, hydrogen, nitrogen, and chlorine), of a large number of elementary substances that were not in the gaseous state at ordinary temperatures (e.g., iodine, carbon, sulfur, calcium, iron, sodium, and barium), and of another group, by analogy, for which the specific gravities were uncertain. The results of all these calculations were presented in four tables that stated, among other things, the specific gravity of the substance and its ratio against hydrogen, which was considered as one. The first table referred to fourteen elements, the second table to combinations with oxygen, the

---

[1] Rocke, *Chemical Atomism in the Nineteenth Century*, 52–53.
[2] Prout, "On the Relation between Specific Gravities of Bodies in the Gaseous State and the Weights of Their Atoms."

From the Atom to Living Systems. Marina Paola Banchetti-Robino and Giovanni Villani, Oxford University Press.
© Oxford University Press 2023. DOI: 10.1093/oso/9780197598900.003.0008

third to compounds with hydrogen, and the fourth, the estimated figures for another twenty-four elements.

There were some discrepancies in the results, which were corrected for the 1816 paper. In this paper, Prout suggested that the atomic weights of all elements were whole-number multiples of the atomic weight of hydrogen, which should therefore be considered the fundamental matter. Prout stated:

> If the views we have ventured to advance be correct, we may almost consider the πρώτη ὕλη [próti ýli or raw material] of the Ancients to be realized in hydrogen; all opinion, by the by, not altogether new. If we actually consider this to be the case, and further consider the specific gravities of bodies in their gaseous state to represent the number of volumes condensed into one; or, in other words, the number of the absolute weight of a single volume of the first matter (πρώτη ὕλη), which they contain, which is extremely probable, multiples in weight must always indicate multiples in volume, and vice versa; and the specific gravities, or absolute weights, of all bodies in a gaseous state, must be multiples of the specific gravity or absolute weight of the first matter (πρώτη ὕλη) because all bodies in their gaseous state which unite with one another unite with reference to their volume.[3]

According to Thomas Thomson, Prout had endeavored to show that the specific gravity of any body could be obtained by multiplying the weights of its atoms by half the specific gravity of oxygen gas; that is, the weight of an atom was always double its specific gravity in the gaseous state (Avogadro's hypothesis). Prout also endeavored to demonstrate that the set of atomic weights of gases could be divided into three groups. In the first group (oxygen, ethylene), the weight of an atom and its specific gravity were represented by the same number. In the second group (chlorine, phosgene, CO_2, nitrogen, steam, etc.), the weight of an atom was double that of the specific gravity. In the third group (HCl, ammonia, HI, etc.), the weight of an atom was four times the specific gravity (volume). "Prout's hypothesis encouraged not only the race for atomic weights but also attempts at classification as a function of atomic weight. Establishing correspondences between the arithmetical ratios of different atomic-weight values and the chemical analogies of elements was like discovering familial relationship and indices of consanguinity. Classifying the elements was like constructing a genealogical tree for the nonliving material world."[4]

---

[3] Prout, "Correction of a Mistake in the Essay on the Relation between Specific Gravities of Bodies in the Gaseous State and the Weights of Their Atoms," 111–113.

[4] Bensaude-Vincent and Stengers, Histoire de la Chimie, 141.125.

Prout is using both atomic weights and molecular weights, thus superimposing two issues. The first issue is that of atomic weights and their correspondence to the weight of hydrogen. The second issue is the number of atoms contained within the molecules of compounds. Although Prout's hypothesis regarding hydrogen as the fundamental matter was eventually proven wrong by Jacob Berzelius (1779–1848) and Jean Servais Stas (1813–1891), his hypothesis regarding atomic weights eventually led to the theory that atomic elements led to the linearization of atomic numbers (protons). By 1869, more elements were discovered, and Dmitry Mendeleev presented his Periodic Table, which will be discussed later in this chapter.

Interestingly, Prout's hypothesis can be considered even more correct today than it was in his own time. If it weren't for the presence of neutrons in atomic nuclei and for the possibility that an element can have multiple isotopes, each with its own atomic weight and percentage in kind, the atomic weight of the elements would effectively be a multiple of, or at least a close approximation of, the weight of hydrogen.

Affinity, Valence, and Molecular Structure

The idea that sympathetic effects originate from an innate relation between two substances and thus from a fundamental affection that cannot be further analyzed had, for a long time, relegated these concepts to a sort of oblivion. Such notions, which had been instrumental to 16th- and 17th-century alchemy and chemistry, constituted attempts to explain how substances were held together. As discussed in a previous chapter, Newton addressed the question of attraction between atoms in Query 31 of his *Opticks*, and, in this text, Newton shows a broad knowledge of chemistry. Like most of his contemporaries, however, Newton does not always make a clear distinction between physical change and chemical reaction. In many respects, Newton's ideas are closely related to the ideas of chemical affinity because affinity can be regarded as similar to his concept of force. However, it is also true that the attempt to subsume the qualitative differences between affinities under a set of general laws was never successful.

Under the hegemony of Newtonian physics, chemists finally began to speculate that the force underlying chemical bonds was nothing other than a manifestation of the gravitational force, although they would make distinctions between his chemical force and his gravitational force. First, the chemical force was considered to be an essential property of matter that was responsible for its intrinsic activity and that was not tied to other properties of matter, such as mass or acceleration. Second, it was conceived as a force that acted via contact rather than at a distance. If such a force is intrinsic and distinct for each element, it can never be conceptualized as a universal law. It was ultimately conceived as being only a force of attraction, so that instances of chemical repulsion were explained as the combination of substances with ethereal fluids. However, the Newtonian theory was still too generic and did not account for the fact that the forces of attraction of each substance were elective and that their behavior changed under the influence of different factors such as temperature, environment, and reagents. Because the gravitational theory did not explain this selectivity of reciprocal action on the part of substances, it was deemed necessary to find another explanation. Additionally, a practical problem was also involved in rendering affinity amenable to Newton's law. The masses of and distances between the planets and the sun were known. However, although the masses of atoms were relatively known, the distances between atoms were completely unknown.

It is in this context, and after reading Newton's description of affinity reactions in the *Opticks*, that Étienne François Geoffroy composed his first tables of affinity in 1718. However, it would not be until 1783 and the work of Lavoisier that the concept of affinity would be clarified as being dependent on temperature. Lavoisier insisted that Geoffroy's tables had been produced at a constant temperature. T. O. Bergman, one of the most fervent adherents of the theory of affinity, held that the chemical affinity of each substance was of a definite and constant strength and that it was also selective, in that it manifested itself only in a unidirectional fashion and only between determinate substances. Following Bergman's work, Guiton de Morveau produced his entry *Chymie* for the 1786 *Encyclopédie méthodique* and distinguished between four different fundamental types of affinity. However, in each case, de Morveau stipulated that the mass of reagents was not to be considered when evaluating the affinity of substances.

The idea of the dependence of affinity on chemical mass, which is a concept analogous to what we today call concentration, was eventually introduced by Claude Louis Berthollet. "Berthollet significantly modified the theory [of affinity] by denying the 'elective' (determinative) characteristic of chemical affinity, replacing it with a more complex and relative version that anticipated modern ideas on chemical equilibrium and mass action. According to Berthollet, affinity was not an absolute force but could be modified by such factors as relative concentration of reactants, solubilities, 'cohesion', and 'elasticity' (precipitation and volatilization of products)"[5] and affinities were proportional to saturation capacities. Joseph Louis Gay-Lussac and other chemists disagreed with Berthollet and pointed out that, empirically, "affinities are found to be *inversely* proportional to saturation capacities."[6]

---

[5] Rocke, *Chemical Atomism in the Nineteenth Century*, 7.
[6] Ibid., 9.

Berthollet's hypothesis did not imply atomism, nor did it make any reference to a particulate theory of matter. Instead, he was "committed to versions of the Newtonian physicalist tradition that sought to discern and quantify the short-range effects of chemical affinity rather than to weigh atoms, a tradition, in other words, that focused on forces rather than on ponderable matter itself."[7] Yet, a crucial step in our understanding of atomic bonding was due to Dalton's recognition that the various symbols used to denote elements could be combined to represent molecules and compounds, in a way that clearly identified the number and type of atoms of which they were composed. The introduction of this notation in which substances could be represented by combinations of atomic symbols implied a persistent interaction between the atoms; that is, it implied bonding atoms.

The several electrochemical theories of affinity that were introduced during the first decades of the 19th century made advancements by reconciling the concept of affinity with chemical atomism. For example, "since 1811, [Jöns Jacob Berzelius] had been urging an electrochemical theory of affinity that traced the chemical difference in compounds to the electrical characteristics of their constituent atoms. This fundamentally materialistic conception seemed distant from—and perhaps inconsistent with—the mechanist-reductionist structural approach, which stressed the arrangements rather than the natures of the ultimate particles."[8] But it was not enough to explain affinity in terms of the electrochemical differences between atoms, since the question remained open as to what enabled atoms to combine according to well-determined numerical relations. In other words, the nature of valence, which is itself intimately linked with a nonreductionist and nonmechanistic concept of chemical structure, was not yet understood.

Edward Frankland (1825–1899) gave the first statement of the modern law of valence in order to explain the results of his experiments producing new organometallic compounds. He stated: "I had not proceeded far in the investigation of these compounds before the facts brought to light began to impress upon me the existence of a fixity in the maximum combining value or capacity of saturation in the metallic elements. . . . It was evidenced that the atoms of zinc, tin, arsenic, antimony &c. had only room, so to speak, for the attachment of a fixed and definite number of the atoms of other elements, or, as I should now express it, of the bonds of other element."[9] Eventually, Frankland recognized that some elements could have multiple valencies. In 1852, he expressed his ideas on valence as follows:

[7] Ibid.
[8] Ibid., 170.
[9] Frankland, *Experimental Researches*, 145, as quoted in Rocke, *Chemical Atomism in the Nineteenth Century*, 249.

When the formulae of inorganic chemical compounds are considered, even a superficial observer is struck with the general symmetry of their construction. . . . Without offering any hypothesis regarding the cause of the symmetrical grouping of atoms, it is sufficiently evident . . . that such a tendency or law prevails, and that, no matter what the character of the uniting atoms may be, the combining power of the attracting element, if I may be allowed the term, is always satisfied by the same number of these atoms.[10]

Here, we may contrapose the concept of affinity with that of valence since affinity focuses on the differences between atoms, while valence focuses on their similarities. An atom is monovalent independently of any other atom to which it may be bonded. Although Frankland and other chemists worked to understand valence more completely, the principal driver for the acceptance of the modern sense of valence was August Kekulé (1829–1896). Although Kekulé's contributions to the theory of valence is intrinsically connected to his work on structural theory, this chapter will focus only on valence, whereas Chapter 8 will discuss Kekulé's concept of structure and his transformational formulas. Kekulé proposed that elements had a fixed valence, but although he was aware of Frankland's work, he did not recognize that combining power was the same as valence.

In a famous paper on chemical compounds, in which he adopts an atomist perspective and links the concept of valence with that of affinity, Kekulé wrote that "[t]he number of atoms of other elements (or radicals) connected with an atom (an element, or, if one does not want to reduce the observation to the elements themselves in the case of composite bodies, a radical) depends on the basicity or relationship size of the components."[11] As Alan Rocke points out, "the major novelties of the paper were twofold: the extension of the principle of 'atomicity' all the way down to the individual atoms; and a detailed discussion of the nature of carbon, including its ability to form bonds to itself."[12] In an 1858 paper on the atomicity of the elements, Kekulé stated:

I consider it necessary, and, in the present state of chemical knowledge, in many cases possible, to go back to the elements themselves which compose compounds, in order to account for the properties of chemical substances. . . . I believe . . . that we must extend our considerations to the constitutions of the radicals themselves, that we must determine the relations of the radicals among

---

[10] Frankland, "On a New Series of Organic Bodies Containing Metals," 440, as quoted in: Rocke, *Chemical Atomism in the Nineteenth Century*, 249–250.

[11] Kekulé, "Über Die, S.G. Gepaarten Verbindungen Und Die Theorie der Mehratomigen Radicale," as quoted in Rocke, *Chemical Atomism in the Nineteenth Century*, 129–150.

[12] Rocke, *Chemical Atomism in the Nineteenth Century*, 266.

each other, and that we must derive the nature of the radicals as well as that of their compounds from the nature of the elements. My earlier considerations on the nature of the elements and the basicity of the atoms serve as the point of departure for these views.[13]

Kekulé's ability to distinguish between atoms of the type and atoms of the radical demonstrates that his theory could successfully clarify how atoms were arranged in a molecule,[14] thereby making important contributions to the notion of organic molecular structure, which will be discussed in Chapter 8. In 1858, A. S. Couper published a paper that reiterated many of Kekulé's views on atomicity and valence, although Kekulé received the larger share of the credit for this new theory of valence. Although organic chemists eventually concluded that each element had a constant valence, the evidence of inorganic chemistry demonstrated that valence is not a constant property, nor is it found in isolated atoms. It depends, instead, on the nature of bonded atoms and the physicochemical conditions in which such atoms interact.

Additionally, valence theories like that of Kekulé-Couper were not able to explain why the "chemical affinity units" of elements involved in reactions are always saturated by the same number of atoms, independently of the chemical character of such atoms. Although the Kekulé-Couper valence theory claimed that all interatomic bonds were the same within molecules, the question of whether such bonds were indeed equal remained open. Valence was critical as a driver for the understanding of organic chemistry, but it was considered a difficult and diffuse concept since, as Peter Ramberg commented, "Strictly speaking, valence . . . was a number that possessed no physical significance."[15] The frustration of 19th-century chemists was expressed by Victor Meyer when he wrote: "The sheer volume of work and the large number of advantages gained have never been able to suppress the awareness that we are currently completely unclear about the basic principle of our current views and the nature of what we call valence or affinity."[16]

This model of fixed valence successfully rationalized the chemistry and the structures of organic compounds using a fixed valence of four for the element of carbon. In addition, it allowed the extension to elements such as nitrogen and oxygen, which were proposed to have fixed valencies of three and to elements such as nitrogen and oxygen, which were proposed to have fixed valencies of three and two, respectively. However, even Kekulé had to accept the reality that some

[13] Kekulé, "Über Die, S.G. Gepaarten Verbindungen Und Die Theorie der Mehratomigen Radicale," as quoted in Rocke, *Chemical Atomism in the Nineteenth Century*, 266–267.

[14] Rocke, *Chemical Atomism in the Nineteenth Century*, 272.

[15] Ramberg, *Chemical Structure, Spatial Arrangement: The Early History of Stereochemistry*.

[16] Meyer, "Zur Valenz Und Verbindungsfähigkeit des Kohlenstoffs," 192–206.

compounds did not fit well in his universal view, and he introduced the description "molecular compounds" for substances that did not conform to his valence rules. He stated: "In addition to these atomic combinations we must distinguish a second category of combinations, which I will refer to as atomic combinations, we must distinguish a second category of combinations, which I will refer to as molecular combinations."[17]

By the latter part of the 19th century, valence or one of its synonyms was being used to refer sometimes to oxidation number and sometimes to the number of bonded atoms. The problem had repercussions for nomenclature, problems that H. G. Madan well summarized in 1869 when he stated: "It is very much to be regretted that the subject of nomenclature is in such an unsettled state. It seems a real reproach to chemists that scarcely two chemical nomenclature is in such an unsettled state and . . . that there [are] hardly two text-books in which the same system of names is adopted."[18]

Mendeleev, the Periodic Table, and Theory of Elements

Although the development of the periodic table of the elements is generally credited to Dmitri Mendeleev (1834–1907), Julius Lothar Meyer (1830–1895) must also be recognized for his contribution to this development.[19] While attending the first-ever international gathering of chemists at the Karlsruhe Congress in 1860, Meyer was introduced to the concept of measuring atomic weights. As a result, his 1864 book *Die modernen Theorien der Chemie* (*Modern Chemical Theory*) examined atomic weights and how they related to the properties of the elements. In this book, Meyer introduced his version of a periodic table that included twenty-eight elements that were split into six families, based on similarities in physical composition, leaving blank spaces for elements that had yet to be discovered.

Meyer's book placed a special emphasis on valence, or combining power, and demonstrated that the atomic weight of the elements increases as integral valence changes. In his original scheme, the valencies of the succeeding families, beginning with the carbon group, were 4,3,1,1,1, and 2. The resulting periodic table that Meyer finally published in 1870, one year after Mendeleev's, did not have the predictive element that ultimately made Mendeleev's contribution more famous. However, Meyer was able to make his own unique substantive contributions to the understanding of periodicity, with graphic demonstrations

---

[17] Kekulé, "Sur L'Atomicité Des Éléments, " 510–514.
[18] Madan, "Remarks on Some Points in the Nomenclature of Salts," 22–28.
[19] Scerri, *Selected Papers on the Periodic Tables*, 54.

of the relationship between atomic volume and atomic weight. Several chemists, including Robert Bunsen, to whom Meyer dedicated his 1864 book, "had doubts about the periodic law at first, but these doubters were gradually convinced by the independent discovery of elements that fit into the blanks in the tabular arrangement and the correction of old atomic weights"[20] that were cast into doubt by Meyer's table.

While working on his own book of chemistry that would ultimately be published in 1871, Mendeleev was on a similar path as Meyer. As he was attempting to organize the elements by writing their properties on cards and arranging them, he realized that, when he put them in order of increasing atomic weight, certain types of elements regularly occurred. For example, a reactive nonmetal was directly followed by a very reactive light metal and then a less reactive light metal. He proceeded to create a tabular system, which he published in 1869, that arranged the elements in horizontal rows, but the system was then changed to the vertical rows that are familiar to us today.[21]

In 1869, Mendeleev ordered the elements according to the established atomic masses, keeping in mind the physical and chemical properties of the substances that were included in his table on the basis of his newly enunciated periodic law. The law stated that "the properties of simple bodies, the constitution of their compounds as well as the properties of the latter, are periodic functions of the atomic weights of the elements, because these properties are themselves the properties of the elements from which these bodies derive."[22] With this emerged the crucial concept of the chemical element as distinct from the simple substance.

> The opposition between "simple" and "compound," which had been the main feature of Lavoisier's system, was pushed into the background for the benefit of a new concept of opposition between the empirical or phenomenal logical reality of simple or compound bodies and the abstract, underlying element. While the "simple substance" was the key concept of a chemistry based on analysis, the element became the key concept and explanatory principle in Mendeleev's chemistry based on substitution. Only the element could explain the properties of simple bodies as well as of combinations. It was the element rather than the simple body that was responsible for the conservation of matter in chemical reactions, the element that circulated and was exchanged.[23]

The law of periodicity Mendeleev cited states that "the properties of simple bodies, the constitution of their compounds as well as the properties of the latter, are

[20] Science History Institute, "Julius Lothar Meyer and Dmitri Ivanovich Mendeleev," 2–3.
[21] Scerri, *Selected Papers on the Periodic Table*, 114.
[22] Mendeleev, as quoted in Bensaude-Vincent and Stengers, *Histoire de la Chimie*, 141.
[23] Bensaude-Vincent and Stengers, *Histoire de la Chimie*, 141.

periodic functions of the atomic weights of the elements, because these properties are themselves the properties of the elements from which these bodies derive."[24]

Mendeleev was not only able to recognize the elements in the correct manner, but if an element appeared in the wrong place due to its atomic weight, he moved it to fit in with the pattern he had discovered. For example, iodine and tellurium should be placed in a different order, based on their atomic weights, but Mendeleev observed that iodine was very similar to the rest of the halogens (fluorine, chlorine, bromine) and that tellurium was similar to the group 6 elements (oxygen, sulfur, selenium), and so he switched their order.[25]

Mendeleev's table had substantive differences from Meyer's, some of which would influence the future recognition of his own periodic system. This is because Mendeleev did not limit himself to simply organizing the known elements but actually predicted as yet undiscovered elements and their properties. As the decades passed, Mendeleev's predictions proved quite accurate. Over the following 15 years after his periodic system was published, three of the predicted elements were discovered. These were Gallium, Scandium, and Germanium.[26] However, the final triumph of Mendeleev's work was the discovery of the noble gases in the 1890s by William Ramsay. Although Ramsay's discovery seemed to initially contradict Mendeleev's predictions, he later realized that they were further proof of his system, fitting in as the final group on the periodic table. This gave the table the periodicity of 8 rather than 7, as it had previously been.[27]

Mendeleev published the work that he believed corresponded best to chemistry. In other words, the work deepened, on the one hand, the understanding of the relationship between composition, reactions, and the amounts of simple and compound bodies and, on the other hand, the intrinsic properties of the elements contained in them. He demonstrated his adherence to atomism when he insisted that the difference between simple bodies and elements was that

[a] simple body is something material . . . with physical properties and capable of chemical reactions. The expression that corresponds to simple body is the idea of the molecule. . . . On the contrary, the name of element characterizes the material particles that form simple and compound bodies and that determine the manner in which these will behave from a chemical and physical point of view. The word element suggests the idea of atom.[28]

[24] Mendeleev (1905), as quoted in Bensaude-Vincent and Stengers, *Histoire de la Chimie*, 141.
[25] Scerri, *The Periodic Table: Its Story and Its Significance*, 84.
[26] Scerri, *Selected Papers on the Periodic Table*, 58.
[27] Scerri, *The Periodic Table: Its Story and Its Significance*.
[28] Bensaude-Vincent and Stengers, *Histoire de la Chimie*, 141.

Mendeleev's concept of the chemical element, which was both a condition and a product of the periodic classification, had a very real and not just a theoretical significance. For Mendeleev, the individuality of the chemical elements was an objective characteristic of nature as fundamental as Newtonian gravitation. "Just as he refused to get embroiled in debates on the reality of atoms . . . he made himself champion of the individuality and the multiplicity of elements and fiercely fought Prout's hypothesis, which he considered a regression into the fantasy world of alchemy."[29]

The atom as a physical entity was constructed from the chemical atom of the periodic table, which showed a complete view of chemistry beginning with its atoms. The table also served as a point of reference, establishing the guideline that could finally convince those chemists who still doubted their physical reality. To the extent that Mendeleev identified chemical elements with atoms and to the extent that he affirmed the individuality and multiplicity of chemical elements, he also affirmed the individuality and multiplicity of physical atoms. However, in affirming the multiplicity of physical atoms, he also introduced the problem of explaining what it is about distinct atoms that makes them distinct.

Mendeleev would have answered that it is mass that makes atoms distinct. However, mass alone cannot fully explain chemical variety since the question that then arises is what accounts for the differences in mass between atoms. This issue would ultimately lead to the idea that atoms themselves are not irreducible but, rather, are structures of even smaller particles. Mendeleev was never satisfied with his lack of understanding of the nature of periodicity, an understanding that would only occur when the composition of the atom was finally understood. He made this point clearly in his 1889 paper:

> Two centuries have elapsed since the theory of gravitation was enunciated. . . . A hundred years later the conception of the elements arose; it made chemistry what it now is; and yet we have advanced as little in our comprehension of simple bodies since the times of Lavoisier and Dalton as we have in our understanding of gravitation. The periodic law of the elements is only 20 years old: it is not surprising therefore that knowing nothing about the causes of gravitation and mass, or about the nature of the elements, we do not comprehend the *rationale* of the periodic law. . . . [However,] the law of periodicity first enabled us to perceive undiscovered elements at a distance which formerly was inaccessible to chemical vision. . . . [But] I unhesitatingly say that although greatly enlarging our vision, even now the periodic law needs further improvements in order that it may become a trustworthy instrument in further discoveries.[30]

---

[29] Ibid., 142.
[30] Mendeleev, "The Periodic Law of the Chemical Elements," 644–645.

It was not until 1913, 6 years after Mendeleev's death, that the final piece of the periodic puzzle fell into place. This occurred when Antonius van den Broek proposed that the atomic number, or nuclear charge, determined the placement of elements in the periodic table. Although he correctly identified the atomic number of all elements up to tin, which is number 50, he did not have a method for experimentally verifying these numbers. As a result, the belief persisted that atomic number was a consequence of atomic weight. It wasn't until Henry Moseley attempted to test van den Broek's hypothesis that he discovered a method for measuring atomic number and an absolute periodic sequence for the elements, as well as discrepancies between atomic number and properties that resulted from arranging elements strictly in terms of atomic mass.[31]

Moseley's experiments involved firing what was, at the time, the newly developed X-ray gun at samples of the elements and measuring the wavelength of the X-rays given. He used this measurement to calculate the frequency and found that, when the square root of the frequency was plotted against the atomic number, the graph showed a perfectly straight line. Within 10 years of his work, the structure of the atom had been finally determined through the work of many prominent scientists. This further explained why Moseley's X-rays corresponded so well with the atomic number.[32]

When an electron falls from a higher-energy level to a lower one, the energy is released as electromagnetic waves, and, in Moseley's experiment, these were X-rays. The amount of energy that is given out depends on how strongly the electrons are attracted to the nucleus. The more protons that an atom has in its nucleus, the more strongly the electrons will be attracted and the more energy will be given out. As we now know, the atomic number is the number of protons, and it is the number of protons that determines the energy of the X-rays.[33]

The discovery of protons and neutrons demonstrated that the classical atom of antiquity and early modern chemistry was an inaccurate concept. As Dalton had established, not all atoms were equal in terms of their mechanistic properties and, as the work of 19th-century chemistry had shown, they were not indivisible. This new knowledge rendered obsolete Lavoisier's definition of chemical elements as the final, albeit provisional, point of analysis. A chemical element came to be defined as a species of atoms with a consistent number of protons, and that number is indicated precisely by the atomic number of that element.[34] Among many other things, this discovery explained the mechanism of several types of radioactive decay, such as alpha decay. With the advent and development of subatomic physics, it was proposed that protons and neutrons themselves were

---

[31] Scerri, *The Periodic Table: Its Story and Its Significance.*
[32] Egdell and Bruton, "Henry Moseley, X-ray Spectroscopy and the Periodic Table."
[33] Ibid.
[34] Scerri, *The Periodic Table: Its Story and Its Significance.*

made of even smaller particles called quarks, which explain the transmutation of neutrons into protons during beta decay.

Although Mendeleev recognized that the mystery of periodicity would not be resolved until the structure of the atom was understood, it would be a mistake to view his theory as a forerunner of 20th-century electron theories. As Bensaude-Vincent and Stengers have stressed:

> [Mendeleev's] periodic system belongs to nineteenth century chemistry, whereas in today's courses and chemistry texts the periodic classification is presented as an expression of the electronic structure of atoms, and Mendeleev, when he is mentioned, is portrayed as a precursor of the electron theories formulated in the twentieth century. Far from being a prophet of future developments in atomic theory, Mendeleev was trying to reorganize the knowledge of his time . . . the unity of chemistry, briefly contested [by the seeming distinction between the organic and the inorganic] seemed to have been rebuilt on a new foundation. The analytic logic reigning in the early nineteenth century had given way to a taxonomic logic of tabulation. Whether organic or inorganic, chemistry obeyed the same laws, and the constituent elements were classified in the same table. The distinction between the two branches seemed justified less by the nature of things than by pedagogical considerations.[35]

Yet, the distinction between the inorganic and organic branches of chemistry was, in fact, more than merely a pedagogical consideration, since the question of what constitutes the difference between nonliving things and living organisms was persistent. It is to this question that we turn in the next chapter, as we examine crucial developments in organic chemistry that advance knowledge from atomic structure, affinity, and valence toward a greater understanding of molecular structure and complexification.

---

[35] Bensaude-Vincent and Stengers, *Histoire de la Chimie*, 143.

8

Organic Chemistry, Molecules, and the Implications for Atomism

As we have seen, the periodic table of the elements, as Mendeleev and other chemists perceived it, demonstrated that all chemistry follows the same laws and that the apparent distinction between inorganic and organic chemistry is merely a pedagogical, rather than a substantive, distinction. However, although the tabular taxonomy of the elements once complete is of course inclusive of all the constituents that make up molecules and compounds in nature, the question of what distinguishes the molecules of living systems from those of nonliving things still remains to be answered. Even if one answers this question by saying that it is the ubiquitous presence of carbon that marks organic molecules and that, in turn, form the cells of living organisms, one must still explain what it is about carbon that makes it possible for ever greater degrees of complexity to emerge from the constituents identified in the periodic table and how we are to understand the nature of systemic complexity. This chapter will begin to address this question by focusing on the development of the molecular structure hypothesis in the 19th century, and the following chapters will trace the further development of the theory of complexification from a chemical perspective.

Background for the Molecular Structure Hypothesis

The molecular structure hypothesis states that a molecule is a collection of atoms linked by a network of bonds. Since the 19th century, this hypothesis has been a successful concept for ordering and classifying the observations made by chemists. The theory of molecular structure is foregrounded by the discovery and study of isomerism and of chemical homology, which suggested that the forces of interaction between the constituent atoms within a molecule are distributed according to a definite order. It would take some time, however, before chemists could determine the distribution of atomic bonds within molecules. Nevertheless, the questions of how atoms are structured within compounds, of how the action of atomic "chemical forces" is distributed within a molecule, and of whether structure and bonding forces influence each other were considered problems of primary importance. The fundamental premises preparing the

From the Atom to Living Systems. Marina Paola Banchetti-Robino and Giovanni Villani, Oxford University Press.
© Oxford University Press 2023. DOI: 10.1093/oso/9780197598900.003.0009

territory for the emergence of structural theory were the theories of valence and molecular bonds, the recognition that carbon atoms could bond in chains, the exact definition of atom and of molecule, and the development of methods for determining precise atomic and molecular weights.

By the beginning of the 19th century, chemists generally had an understanding that a single chemical formula represented a single compound, and, "with the notation of Dalton and Berzelius in place, the constitution of compounds could be expressed by the combination of atomic symbols."[1] In 1824, Justus von Liebig and Joseph Gay-Lussac described the preparation and properties of silver fulmi-nate, AgCNO, which was a compound with interesting properties: "Silver ful-minate never detonates alone at the temperature of 100° or 130°, but it should not be exposed to the slightest shock between two hard bodies, even when it is in water."[2] In the same year, Friedrich Wöhler described silver cyanate (cyanure silver oxide) and reported the analysis AgOCN—in other words, the formula-tion identical to Liebig's silver fulminate.[3]

After a few exchanges between Liebig and Wöhler, Liebig eventually conceded that the two compounds had an identical formulation but different properties.[4] In 1828, Wöhler proceeded to show that syntheses designed to yield ammonium cyanate, NH_4OCN, gave the isomeric organic compound urea, $OC(NH2)2$.[5] "The difference between silver fulminate and silver cyanate, with the modern formulations AgCNO and AgOCN, respectively, leads irrevocably to the need for the concept of chemical bonding and for the consequence that atoms have fixed interactions with specific other atoms and ultimately that the bonding of atoms implies a three-dimensional arrangement in space."[6]

It was Berzelius who finally took an array of disparate results and correlated them into one of the most important insights in the development of modern chemistry.[7] Among these results were the pairs of compounds obtained by Liebig, Gay-Lussac, and Wöhler. In 1830, Berzelius "proposed that these pairs of compounds should be described as homosynthetic or isomeric (Greek, ἰσομερής, equal parts) bodies,"[8] that is, compounds having the same chemical constitu-tion and molecular weight but having different chemical properties.[9] With the

[1] Constable and Housecroft. "Chemical Bonding: The Journey from Miniature Hooks to Density Functional Theory," 15.
[2] Liebig and Gay-Lussac, "Analyse du fulminate d'argent," 285–311.
[3] Wöhler, "Analytische Versuche Über Die Cyansäure," 117–124.
[4] Cohen and Cohen, "Wöhler's Synthesis of Urea: How Do the Textbooks Report It," 883.
[5] Tóth, "Demonstration of Wöhler's Experiment: Preparation of Urea from Ammonium Chloride and Potassium Cyanate," 53.
[6] Constable and Housecroft, "Chemical Bonding: The Journey from Miniature Hooks to Density Functional Theory," 16.
[7] Ibid.
[8] Ibid.
[9] Ramberg, "The Death of Vitalism and the Birth of Organic Chemistry: Wöhler's Urea Synthesis and the Disciplinary Identity of Organic Chemistry."

insights of Berzelius, the stage was set for the growth of organic chemistry into a rigid intellectual discipline. In this regard, two perspectives regarding the nature of molecular structure were contraposed.[10] These two distinct perspectives will be the focus of this chapter.

On one side was August Kekulé's assertion that rational formulae are transformation formulae, rather than structural formulae. In other words, rational formulae could not be used in any way to express structure, that is, the position of atoms within compounds. This view was based on the idea that the original structure of atoms within a molecule could not be extrapolated through the separation of atoms from molecules that were either decomposing or undergoing other changes.

On the other side was the view expressed by the Russian school, represented by Alexander Butlerov (1828–1886), who wrote in 1861 that

> conclusions about the chemical structure of substances can be plausibly based on the study of their formation through synthesis and, primarily, through those syntheses that occur at low temperatures and under conditions that allow chemists to follow the gradual development of complex relations between chemical particles. Only one rational formula will be possible for each substance and, once we know all the general laws that describe the dependence of chemical properties on chemical structure, such a formula will express all of the properties of a substance.[11]

Butlerov was also one of the pioneer chemists who depicted organic molecules with a two-dimensional structure connected by lines, which represent the bonds or valencies.[12] As Butlerov wrote, "Starting from the assumption that each chemical atom possesses only a definite and limited amount of chemical force (affinity) with which it takes part in forming a compound, I might call this chemical arrangement, or the type and manner of the reciprocal binding [*gegenseitingen Bindung*] of the atoms in a compound substance, by the name of chemical structure."[13] The term *gegenseitig* "indicates that the chemical bond is the outcome of an action carried out by both atoms involved in it."[14] These developments can be seen as marking the beginning of the structural theory of organic chemistry.

---

[10] Rocke, "The Reception of Chemical Atomism in Germany."
[11] Butlerov, as cited in: Minkin, "Current Trends in the Development of A. M. Butlerov's "Theory of Chemical Structure," 1265–1290.
[12] Butlerov, "On the Different Explanations of Certain Cases of Isomerism," 9–12.
[13] Minkin, "Current Trends in the Development of A. M. Butlerov's Theory of Chemical Structure."
[14] Ghibaudi, Cerruti, and Villani, "Structure, Shape, Topology: Entangled Concepts in Molecular Chemistry," 295.

Kekulé's Concept of Transformation Formulae

The concept of valence is critical for a true understanding of modern organic chemistry, and the principal driver for the acceptance of this concept was August Kekulé (1829–1896).[15] Kekulé's model of fixed valence for atoms was successful in rationalizing the chemistry and structures of organic compounds using a fixed valence of four for carbon. This model allowed for the extension of fixed valence to elements such as nitrogen and oxygen, which were proposed to have fixed valencies of three and two, respectively. One of the successes of this model was that the "fixed valence of four for carbon is necessitated by the presence of multiple bonds (or free valences)"[16], although Kekulé had to admit that some compounds did not fit well in his universal view.

He introduced the description of molecular compounds for substances that did not conform to his rules of valence. He stated that "[i]n addition to these atomic combinations we must distinguish a second category of combinations, which I will refer to as molecular combinations."[17] Kekulé's terms *atomic valence* and *molecular valence* refer, respectively, to the principal valence that is saturated in the molecular structure and the secondary valence. Today, we understand that there exists a large class of chemical compounds, referred to as "complex," that are formed after the valencies have been saturated in the molecular structure. These are formed with bonds between molecules, and the simplest example is the ion NH_4+, which is formed when $H+$ is added to NH_3 (ammonia), utilizing the coordinate bond of ammonia. Regarding compounds, Kekulé wrote:

> I consider it necessary, and, in the present state of chemical knowledge, in many cases possible, to go back to the elements themselves which make up compounds, in order to account for the properties of chemical substances. I no longer consider the principal task of the times to be the determination of atomic groupings which (due to certain properties) can be considered as radicals, thus assigning compounds to a few types.... On the contrary, I believe that we must extend our considerations to the constitutions of the radicals themselves, that we must determine the relations of the radicals among each other, and that we must derive the nature of the radicals as well as that of their compounds from the nature of the elements.[18]

---

[15] Palmer, *A History of the Concept of Valency to 1930*.

[16] Constable and Housecroft. "Chemical Bonding: The Journey from Miniature Hooks to Density Functional Theory," 13.

[17] Kekulé, "Sur l'atomicité des éléments," 510–514.

[18] Kekulé, "Ueber die s.g. gepaarten Verbindungen und die Theorie der mehratomigen Radicale," 109–131.

"Kekulé then developed in detail the consequences of regarding carbon as tetravalent and of the ability of tetravalent carbon atoms to form C-C bonds with each other. This step was crucial and unprecedented in the development of organic chemistry. To use Kekulé's language, [he was able to fully generalize] the theory of polyatomic radicals, transforming it into the theory of the atomicity of the elements."[19] Rocke tells us that

Kekulé was by no means the first chemist to conceptualize the constitutions of organic compounds based on the atomicity of the elements. However, he was the first to inspire a continuing theoretical and experimental tradition based on this idea. In this sense, his work formed a striking parallel to that of Rudolf Clausius and others, who were also writing in the late 1850s and who established a tradition leading to a general acceptance of the kinetic theory of gases. Chemists and physicists were simultaneously, and for the first time, exploring the submicroscopic realm in a way that eventually proved empirically satisfying to the majority of scientists in their respective disciplines.[20]

In his *Lehrbuch der organischen Chemie* of 1859, Kekulé clarified his discussion of organic constitutions by distinguishing between atoms "of the radical" (the carbon skeleton itself and all atoms directly and completely bound to it) and atoms "of the type" (atoms indirectly or incompletely bound to the carbon skeleton).[21] Thus, Kekulé was able to make what were at the time quite subtle structural distinctions, for example, between the three oxygen atoms of glycolic or lactic acids.[22] Kekulé reiterated his viewpoint that both radicals and types were merely relative and conventional, as well as mutually complementary, concepts. Different reagents could act upon the same molecule in different ways, revealing different aspects of its constitution. Accordingly, it was clear that a substance could be conceived as belonging to more than one type or as containing different constituent radicals.

A single reaction disclosed a single such type-radical relationship and could be summarized in a rational formula, that is, a formula suggesting the atomic groupings revealed by chemical reactions. Various different formulae could be used for any given substance, depending on the specific reaction or property of the substance that one happened to be discussing.[23] Fully resolved rational formulae were useful when explicitly discussing theoretical matters but should otherwise be avoided as unnecessarily complex. Kekulé emphasized that his graphic formulae were precisely equivalent to his completely resolved

[19] Rocke, "Kekulé, Butlerov, and the Historiography of the Theory of Chemical Structure," 35.
[20] Ibid.
[21] Kekulé, "Ueber die s.g. gepaarten Verbindungen und die Theorie der mehratomigen Radicale," 109–131.
[22] Rocke, "Kekulé, Butlerov, and the Historiography of the Theory of Chemical Structure," 35.
[23] Ibid.

type-formulae. Both kinds of formulae expressed, in the most complete form, the theoretical ("rational") conceptions that one may have of the constitution of a molecule.[24]

Kekulé distinguished his conception of provisional rational formulae derived from chemical reactions from the view that absolute constitutional formulae expressing the actual groupings of the atoms were the proper goal of chemical theory. Two factors made this goal problematic. First, chemical knowledge was derived not from the physical properties of unchanging molecules but from the chemical properties of molecules undergoing sometimes major alterations. This made any inferences regarding the constitutions of the molecules before the reaction uncertain. Kekulé wrote that "[r]ational formulae are reaction formulae and can be nothing else in the present state of science."[25] Second, even if a conception of spatial arrangement could be gained, no notation had yet been devised that could accurately depict three-dimensional atomic arrangements on a two-dimensional piece of paper.[26] Ironically, Kekulé himself was to do much work after 1858 to clarify spatial groupings of atoms by, for example, stipulating rules of inference to be used in determining constitutions from the study of chemical reactions and even devising various models and notational schemes for expressing three-dimensional structures.[27]

"Kekulé thus distinguished between chemical and physical atoms and arrangements or, more precisely, between the apparent atomic arrangements deduced from chemical properties and the true physical (or spatial) arrangement of the atoms within a molecule."[28] Thus, Kekulé affirmed a distinction, which had also been stressed by Gerhardt, between the apparent atomic arrangements deduced from chemical properties ("chemical constitution" or, later, "chemical structure") and the true, actual spatial arrangement of atoms within a molecule.

Kekulé, however, did not deny the possibility that positive science might, in principle, determine the actual arrangement of atoms. But he stressed that this question was independent of the treatment of rational formulae as relative, conventional, and based solely on chemical reactions. Kekulé's theories "required no specific arrangement in space but did refer to chemical relations between atoms in three dimensions."[29] However, as the 1860s evolved, Kekulé's ideas were shifting more and more to atoms having a distinct spatial as well as a valence relationship. In the models he used to represent molecules, a saturated carbon was represented by a tetrahedron. By 1865, Kekulé had proposed the hexagonal

[24] Ibid.
[25] Ibid.
[26] Ibid.
[27] Ramsay, "Molecules in Three Dimensions," 6–11.
[28] Rocke, "Kekulé, Butlerov, and the Historiography of the Theory of Chemical Structure," 33.
[29] Hein, "Kekulé and the Architecture of Molecules," 1–12.

structure for benzene for which he is now famous and had also identified different isomers of the disubstituted derivatives.[30]

Beginning in the middle of the 19th century, "bonds were routinely being represented with the lines that we are familiar with today, although it is also fair to say that there was little understanding or consensus as to what the lines actually meant. The prevalence of the Kekulé view and its emphasis on a fixed valence of four for carbon was instrumental in this development."[31] Some debate arose regarding whether the depiction of the bonds implied anything about the spatial arrangement of the atoms. Slowly, the organic chemistry community came to accept Kekulé's viewpoint, with observations relating to isomerization and the need for a fixed three-dimensional structure.[32]

Atomic Bonding and Butlerov's Structural Theory

It is worthwhile, at this point, to examine how bonds were interpreted in the general organic chemistry community at this time and, for our discussion, we shall make substantive use of the significant research done by Alan J. Rocke on the development of the concept of bond in the 19th century. As the structural theory developed, the concept of bonding had a topological meaning in terms of how the atoms were connected. However, chemists' general understanding was that the depictions that correctly showed the connectivity did not imply anything regarding the positions of the atoms in space.[33]

Although the notion of chemical structure had been used prior to the work of Alexander Butlerov, he gave it a new meaning by including in it the disposition of interatomic bonds. Butlerov's central point was his firm opinion that the concept of atomicity could lead to a perfectly general and perfectly consistent theory of the constitution of chemical compounds. He expressed these ideas as follows:

> Together with Gerhardt we deny for the present the possibility of accounting for the positions of the atoms in the interior of a molecule; it seems quite obvious that chemistry, which only deals with bodies in a state of transformation, is powerless to judge this mechanical structure, as long as physical investigations are not brought to bear on the question. On the other hand, however . . . I am sure that no one would say that this will remain inaccessible to us even in the future. To be sure, we do not know what connexion exists between

---

[30] Kekulé, "Sur la constitution des substances aromatiques," 98–110.
[31] Constable and Housecroft. "Chemical Bonding: The Journey from Miniature Hooks to Density Functional Theory," 17.
[32] Ramberg, *Chemical Structure, Spatial Arrangement. The Early History of Stereochemistry.*
[33] Rocke, "Kekulé, Butlerov, and the Historiography of the Theory of Chemical Structure."

the relative chemical effect of the atoms inside a compound molecule and their relative mechanical positions; we do not even know whether, in such a molecule, two atoms which directly affect each other chemically are in fact situated next to one another, but we cannot deny, putting the concept of physical atoms entirely to the side, that the chemical properties of a body are determined in particular by the chemical bonding [*Zusammenhang*] of the elements which form it. Proceeding from the assumption that there inheres in each chemical atom only a specific limited quantity of chemical force (affinity), with which it participates in the formation of bodies, I would designate this chemical cohesion, or the manner of mutual bonding of the atoms in a compound body, by the name chemical structure. The familiar rule which says that the nature of a molecule is determined by the nature, quantity, and arrangement of its elementary components could thus temporarily be altered in the following way: the chemical nature of a molecule is determined by the nature and quantity of its elementary components and by its chemical structure.[34]

Rocke clarifies Butlerov's point by stating that, according to Butlerov,

we cannot know the actual spatial arrangements of the atoms in a molecule, its 'mechanical structure', at least not by the study of chemical reactions alone. We can, however, investigate the apparent arrangement as revealed by chemical reactions, without having to make any judgment as to whether this is identical to, or even related to, the actual arrangement. To clarify this epistemological distinction, Butlerov spoke of chemical nature, chemical atoms, chemical force, chemical cohesion, and chemical structure.[35]

It was possible to investigate chemical structure by means of chemical experiment alone, without being troubled by questions that were not yet answerable. The distinction that lay at the base of Butlerov's conception of chemical structure derived, as he noted, from Gerhardt. Butlerov had stressed it 3 years earlier, and Kekulé had done so several months before that.[36]

Butlerov suggested that his conception of chemical structure would lead to formulae that are 'truly rational', so that only one such absolute formula would be possible for any given substance. He insisted that type formulae could be used to express chemical structures, without implying that the type *theory* was thereby advocated . . . However, [type formulae] had two disadvantages: chemists would erroneously associate these structurally intended type formulae with the

[34] Butlerov, "Einiges uber die chemische Structur der Korper," as cited in Rocke, "Kekulé, Butlerov, and the Historiography of the Theory of Chemical Structure," 35.

[35] Rocke, *Origins of the Structural Theory in Organic Chemistry*, Volume 2, 357.

[36] Rocke, "Kekulé, Butlerov, and the Historiography of the Theory of Chemical Structure," 35.

non-structural type theory and this notation required an unacceptable amount of space, especially for more complicated molecules.[37]

Butlerov asserted that chemists should embrace a single principle, atomicity, and he was confident that the comprehensive acceptance of this principle would reveal the "*chemical* relationships of the chemical atoms, the chemical bonding arrangement, and would ultimately result in a full understanding of the chemical structure."[38]

In an 1861 speech, Butlerov defined chemical structure as "chemical cohesion, or the manner of mutual bonding of the atoms in a compound." In his 1863 paper on isomerism, he defined it as "[t]he manner of . . . the chemical cohesion [*Zusammenhang*] of the individual atoms forming a complex body."[39] In an 1864 publication, he defined chemical structure as "[t]he sequence [*Reihenfolge*] of mutual action—the manner of the mutual chemical bonding of the elementary atoms in a molecule."[40] In sum, Butlerov denied chemical structure "as the way atoms are chemically connected together to form molecules."[41] Rocke argues that Butlerov "intended chemical structure "to mean no more, and no less, *than the consistent application of valency rules in the schematic construction of molecules*, with an added epistemological caution regarding chemical versus physical knowledge."[42]

If this interpretation is correct, Butlerov's notion of chemical structure seems to be a reformulation of Kekulé's theory of the atomicity of the elements. This similarity is borne out by Butlerov's 1861 definition of "chemical structure." He stated that "[t]he idea of chemical structure arises directly from the concept of valency."[43] "Butlerov stressed that it was not sufficient to consider merely the atomicities of atoms forming a molecule and its type. Rather, it was necessary to determine the precise way in which the atoms were chemically bonded to one another. He was also issuing a call for greater consistency and completeness in pursuing all consequences of the doctrine of atomicity."[44]

Butlerov affirmed forcefully and convincingly the possibility and necessity of determining chemical structures by thoroughly applying the concept of atomicity, purified of all earlier associations and misleading doctrines. He emphasized the need to apply "comprehensively and consistently those ideas that

[37] Ibid., 36.
[38] Ibid.
[39] Butlerov, "Uber die verschiedeneñ Erklarungsweisen einiger Falle von Isomerie," as cited in: Rocke, "Kekulé, Butlerov, and the Historiography of the Theory of Chemical Structure," 35.
[40] Butlerov, "Vvedenie k polnomu izucheniiu organicheskoi khimii," as cited in: Rocke, "Kekulé, Butlerov, and the Historiography of the Theory of Chemical Structure," 35.
[41] Rocke, "Kekulé, Butlerov, and the Historiography of the Theory of Chemical Structure," 37.
[42] Ibid.
[43] Butlerov, "Uber die verschiedenen Erklarungsweisen einiger Falle von Isomerie," as cited in: Rocke, "Kekulé, Butlerov, and the Historiography of the Theory of Chemical Structure," 35.
[44] Rocke, "Kekulé, Butlerov, and the Historiography of the Theory of Chemical Structure," 37.

had been championed by earlier chemists but that had not yet caught hold in the chemical world."[45] After 1861, there was a growing acceptance and appreciation of structural theory, and, in an 1861 speech, Butlerov concluded as follows:

> I am far from thinking that I am proposing a new theory here; on the contrary, I believe that I have expressed views to which very many chemists subscribe. . . . My remarks today were only intended to suggest that the time has come to apply the idea of atomicity and of chemical structure in every case, and entirely free from the type viewpoint, as a foundation for the consideration of chemical constitution, and that this would seem to provide a means of helping chemistry from its present uncomfortable position.[46]

Actually, Butlerov simply maintained that he had contributed to the development of structure theory, not to its origin. Like Kekulé and Gerhardt, he never believed that actual atomic arrangements could be determined.[47] As he remarked:

> Although such formulae do not represent constitutional formulae, as long as constitution is defined—by Gerhardt and Kekulé—as the true arrangement of the atoms, nevertheless many of Kekulé's formulae are clearly intended to express the actual points of attack of affinity. . . . In [Hermann] Kolbe, as in Kekulé, one sees the same tendency to judge the chemical cohesion of the individual atoms forming a complex molecule, or the points of attack of chemical affinity, which is saying the same thing. Kolbe defines constitution as the manner of this cohesion; I call this chemical structure, and prefer this expression because unlike the word 'constitution', it has not been used to mean various different things, and so it can lead to fewer misunderstandings. . . . Be that as it may, most of Kekulé's as well as Kolbe's formulae are clearly based on the principle of chemical structure. I might add that ever since the importance and role of atomicity was recognized, this principle has formed the true point of departure for almost all theoretical considerations.[48]

Although Butlerov recognized that both Kekulé and Kolbe adhered to the theory of structure, he criticized many of Kekulé's specific formulations, of which several contradicted the principle of atomicity, according to Butlerov. He also criticized some of Kekulé's formulations for being incompletely resolved and

---

[45] Ibid.
[46] Bykov, "Origin of the Theory of Chemical Structure," 222.
[47] Rocke, "Kekulé, Butlerov, and the Historiography of the Theory of Chemical Structure," 38.
[48] Butlerov, "Uber die verschiedenen Erklärungsweisen einiger Falle von Isomerie," as cited in: Rocke, "Kekulé, Butlerov, and the Historiography of the Theory of Chemical Structure," 35.

mentioned two instances in which Kekulé had suggested two different chemical structures for the same substance.[49]

Butlerov granted that a 'truly rational' formula might be expressed in various fashions, showing more or less complete resolution of structure, but the structure itself must be unambiguous and must not contradict the theory of atomicity. Kekulé occasionally transgressed this principle. And Butlerov discussed the general application of the principle of chemical structure to the explanation of isomerism, a topic covered earlier by Kekulé, but without special emphasis and with occasional obscurities and inconsistencies.[50]

Butlerov also criticized the concept of topographic arrangement, since he believed that it suggested the actual spatial positions of physical atoms.[51]

In his 1864 reply to Butlerov, Kekulé emphasized that he had come to reject the term *reaction formulae*, adding that "[o]ne sort of rational formula has been given preference in this textbook: namely that sort which includes views concerning the manner of bonding of the atoms composing the molecule, views which derive from the theory of the atomicity of the elements."[52] Here, Kekulé explicitly discussed his definitions of atoms "of the type" and atoms "of the radical," which had informed his theory since at least 1858.[53] "His rational formulae based on these definitions directly indicated: (1) the carbon skeleton; (2) those atoms assumed to be bonded completely to the skeleton (such as hydrogen or carbonyl oxygen); (3) those atoms assumed to be only partially bonded to carbon (such as hydroxyl oxygen); and (4) those atoms thought to be only indirectly combined with the carbon chain (such as hydroxyl hydrogen)."[54]

As he had in 1859, Kekulé emphasized

that fully resolved rational formulae were equivalent to his graphic formulae, that he had always regarded the latter as expressing 'most clearly and completely' the manner of bonding within the molecule, that only one graphic formula was possible for a given substance but that various less fully resolved formulae might be used, depending on the context, as long as they were consistent with the theory of atomicity. This was a position with which Butlerov had no quarrel. It constituted a clarification of some, but not all, of the obscurities and inconsistencies exposed by Butlerov.[55]

An important point regarding the concept of structure, and one that had already been suggested in the 19th-century notion of molecular structure, must be

---

[49] Rocke, "Kekulé, Butlerov, and the Historiography of the Theory of Chemical Structure," 38–39.
[50] Ibid., 39.
[51] Ibid.
[52] Kekulé, *Lehrbuch der organischen Chemie*, as cited in Rocke, "Kekulé, Butlerov, and the Historiography of the Theory of Chemical Structure," 39.
[53] Rocke, "Kekulé, Butlerov, and the Historiography of the Theory of Chemical Structure," 39.
[54] Ibid.
[55] Ibid.

highlighted. This is the point, made by Butlerov in the following passage and that we later take up again in a later chapter in our discussion of atoms in molecules. In 1864, Butlerov developed his theory on the reciprocal action of atoms involved in the composition of a given molecule. He stated that two atoms that are identical in their nature will acquire a different chemical character when they both enter into the composition of a molecule, since the influence that each atom exercises upon the other component parts of this molecule is different from the influence that is exercised by the other atoms.[56]

The work of Vladimir Markovnikov in the 1870s further contributed to the development of structural theory. Markovnikov understood that the affinities or, as we would say today, the "chemical bonds" that an atom can form in its free state were different from those that an atom displays when it is part of a molecule. He understood that, within the molecule, the reciprocal actions can either weaken or strengthen the affinities (or bonds) between atoms. Such actions would diminish with an increasing distance in the chain between the two atoms. Markovnikov stated that the longer the chain forming a molecule, the weaker the reciprocal influence between its members due to the distance that separated them.

The introduction of the theory of molecular structure radically altered organic chemistry. Instead of engaging in blind attempts during the synthesis of new compounds, the new approach followed a method that was based on knowledge of the structure of the starting products and of those being synthesized. Although the theory of molecular structure could not predict the existence of optical and geometric isomers, the development of stereochemistry did not undermine the theory of molecular structure but, rather, gave it greater depth. Successful syntheses succeeded in confirming the fertility of the theory molecular structure, and its validity has never since been called into doubt.

Replacement of the concept of entity with that of molecule (structured entity) not only modified chemistry but also profoundly changed the conception of the microscopic world. For chemists, the material world had always been formed by a large variety of different substances to which they attributed distinct names. The concept of molecular structure corroborated these differences by determining that there is a distinctive molecular structure for each type of molecule. However, the difficulty of explaining what held atoms together remained open. To overcome this difficulty, it became necessary to "enter" into the atom itself and to discover that it was not an elementary entity at all but was itself also structured. These insights will be examined in greater detail in Chapter 10 of this book.

---

[56] Butlerov, "On the Different Explanations of Certain Cases of Isomerism."

9

The Relationship between Chemistry and Biology in the 19th Century

Biology as we understand it today evolved substantially in the 19th century, and its relationship to the chemistry of that century is very complex. Without any pretense of exhaustiveness, we will develop this relationship along three lines: the material, the energetic, and the entropic. By the material direction, we mean the one that deals in the broad sense with "living matter" and with the "processes" in which it is involved. The philosophical problem of vitalism falls within this discussion, and, in this regard, we will be mainly concerned with three historical moments: (1) the artificial synthesis of urea by Friedrich Wöhler, (2) the dispute between Pierre-Eugène-Marcellin Berthelot and Louis Pasteur, and (3) Eduard Buchner's discovery of cell-free fermentation. With regard to the energetic and entropic directions, we will analyze the birth of thermodynamics and its influence on the relationship between chemistry and biology. Justus von Liebig's description of animals as "walking stoves" and Claude Bernard's concept of *milieu intérieur* will be considered as part of the energy field. Finally, the entropic discussion will also consider the general problem of reconciling the increase of entropy (and of "disorder") that is linked to the Second Law of Thermodynamics, showing evidence that the living world is organized/ordered and its species, evolving then becoming extinct, lead to/determine an evolution toward increasing complexity.

Matter and the Processes of Living Beings

One of the "historical" problems of biology, as the study of the animate and its relationship with the sciences of the inanimate, was always whether or not there was a material difference between these two worlds. We can say that the history of biology, understood as the study of the living, essentially embraced two main conceptions of life. In one conception, the substance of living beings is considered special and unique, and, in the other conception, the matter of living beings is considered identical to that of the nonliving, although its organization is considered special and unique.

From the Atom to Living Systems. Marina Paola Banchetti-Robino and Giovanni Villani, Oxford University Press.
© Oxford University Press 2023. DOI: 10.1093/oso/9780197598900.003.0010

For Aristotle, living beings and the world of inanimate objects were composed of the same natural elements and the same matter. However, it is important to distinguish Aristotelian materialism from the 17th-century mechanistic conception of matter that was endorsed by Descartes and his followers. Although Aristotle unites matter and form in a manner that can be justifiably classified as materialist, his cosmology was teleological and holistic. On the other hand and as we shall see later, the 17th-century mechanistic conception of matter was explicitly anti-Aristotelian, antiholistic, and antiteleological.

For Aristotle, the quality that was intrinsic to life was therefore not to be sought in matter but in its form, as understood in the Aristotelian sense. All beings were the result of the union of a matter with a form, and in the case of the living this meant the union of the body with the soul. For Aristotle, the soul was the form of the body. The type of soul present in a living being determined the function of this being. With regard to human beings, Aristotle believed there were three types of soul: (1) the nutritive soul, which corresponded to human beings, other animals, and plants, (2) the sensitive-locomotive soul, which corresponded to human beings and other animals, and (3) the rational soul, which was uniquely human. Although Aristotle unites the soul to the body in the sense that the soul, as the form of the body, cannot exist separately from the body, he does not believe that soul and body are one and the same substance.

The history of biology from the ancient Greeks to the 17th century has been characterized by the opposition of animism and strict materialism. Although materialism is not necessarily mechanistic, it became explicitly associated with the mechanistic cosmology in the 17th century. Thomas Hobbes is one of the first philosophers to offer a strictly materialist and mechanistic cosmology in his *Leviathan* (1651) and to work out its implications for psychology, ethics, and political philosophy. The opposition of animism and materialism no longer appeared openly beyond the 18th century but it continued to underlie subsequent explanations. This remains a fundamental problem, which persists even when the soul has disappeared from philosophical/scientific concepts.

René Descartes is the philosopher most often associated with articulation of the mechanistic theory of matter, which was the culmination of the anti-Aristotelian, antiteleological, and antiholistic cosmology that defined the Galilean scientific revolution. In his *Traité de l'homme* (1664), Descartes postulates that human beings have both a soul and a body, that he wants to separate the study of the one and of the other, and then show how these two components connect. Strictly speaking, for Descartes, there is no such thing as a "living body." All bodies are material and, in accordance with the mechanistic theory, all matter is inert and "lifeless." Thus the human body, like all material bodies, is nothing more than a machine governed by the laws of mechanics, which has heat for the engine, "a fire without light" placed in the heart. However, and unlike all other bodies, the human body is "inhabited" by a nonmaterial substance that we may call "soul" or, more precisely, "mind." Descartes refers to matter as *res extensa* (extended substance) and to mind as *res cogitans* (thinking substance), thereby specifying the essential distinction between these two substances. For Descartes, it is the presence of *res cogitans* that renders human beings both conscious and alive. Thus, since he claims that *res cogitans* is unique to human beings, Descartes does not believe that other animals are either conscious or even alive. Human beings are the only living beings, "the ghost in the machine," while all other beings are simply machines.

Although we as authors of this book clearly disagree with Descartes' assessment regarding other living beings, we must clarify that Descartes is applying the explanatory principle of Ockam's Razor, which as noted earlier, bids us to provide the simplest explanation possible for all phenomena and to avoid postulating superfluous substances or principles. Descartes argues that the method of introspection reveals the presence, in human beings, of subjectivity, voluntary movements of the body, and rational thought; these, he states, cannot be explained by strict mechanistic materialism and thus require the postulation of *res cogitans*—conscious, living, and thinking substance. However, the movements of all other material bodies, including those of what we call animals, can be fully explained via the mechanistic laws that govern *res extensa* and do not require the postulation of *res cogitans* in those bodies.

For Descartes, *res extensa* (matter) and *res cogitans* (soul or mind) are two radically distinct substances that are defined according to entirely opposing essential properties. *Res extensa* is defined as extended substance and is governed by the laws of mechanics, while *res cogitans* is defined as thinking substance and is governed by the laws of logic. In human beings, although the material body's involuntary actions are governed by mechanistic principles, voluntary actions (that is, willed actions) are directed by the mind. In addition, sensation and perception are the processes whereby the mind becomes aware of the body's movements in space and interactions with other material bodies. Thus, although matter and mind are entirely distinct substances that are governed by entirely distinct laws, they do interact in the context of perception and voluntary action.

According to Descartes, this interaction takes place in the pineal gland of the brain, which is where he locates the presence of the mind. It is from its location in the pineal gland that the mind both receives sensations and perceptions from the body and governs the body's voluntary actions. Descartes chooses the pineal gland as the location of the mind because, since the mind is unified and indivisible (because it is not extended), it can only be located in that part of the brain that is itself unified and undivided. This is the pineal gland, which, unlike other parts of the brain, is not divided into hemispheres.

Clearly, the Cartesian conception of life was strictly limited to human life, and a science of biology, from the Cartesian perspective, restricted itself to the study of that aspect of human beings that makes us alive—that is, the mind. Thus, for all practical purposes, biology would be reducible to psychology (the study of subjective experience) and to logic (the study of the rules that govern thought). In this regard, Descartes did not really make any significant contributions to the development of biology as a science. Biology could only advance if at least some aspects of matter could be considered as active and potentially alive. In this regard, the chemical aspects of matter would be crucial to the development of biology. The history of biology after Descartes can, therefore, be considered as the way in which the mechanical vision of "passive" matter and the chemical vision of "active" matter have been reconciled.

Beginning from passive matter, biology was transformed into a conception of matter with different physical properties (gravitation, electricity, magnetism) and, above all, different chemical properties and activities independent of mechanical organization. These chemical activities were at first ill-defined, in very much the same way as vital activities had been. They then became more and more clearly defined, which led to living beings no longer being considered as mechanical machines but rather as chemical machines. Therefore, the problem became that of the relationship between the mechanical organization (involving the essentially spatial and nonintrinsic quality of matter) and the chemical or inherent qualities and activities, with all the gradations between mechanical organization and chemical activity having to be considered. With regard to living beings, the mechanical approach was articulated with reference to the organs, while the chemical approach was articulated with reference to their functions. The evolution from mechanics to chemistry is well exemplified in the evolution of physiology. The chemical evolution begins with a physiology of the organs, which focuses on their structure, and is followed by a physiology of the functions, which focuses on the chemical activities that take place in these organs. Some activities are relatively independent of the organ structure and can occur in very different structures.

Newtonian physics dominated the late 17th, 18th, and 19th centuries and became the new science of reference that replaced Galilean mechanics. Therefore, it influenced biology's image of the animal-machine, which was created by Cartesian mechanistic philosophy on the basis of Galilean physics. Newton's most important influence in the context of biological explanation is related to his conception of gravitational attraction. At this point, this question is no longer reduced to simple inert extension but includes the concept of activity. Newton does not say that gravity is an intrinsic property of matter because he is careful not to specify the nature and cause of gravitational attraction. But since Newton, attraction has been considered an intrinsic quality of matter, one of its essential qualities in the same way as extension, impenetrability, or inertia.

Since Newton, the chemical and mechanical approaches have always had to take each other into account. These two paths converge toward an approach to life that takes into account both mechanical organization and chemical activities inherent to matter. Therefore, the animal-mechanical machine becomes the animal-mechanical-chemical machine. However, the difficulty was, and remains, how to reconcile these two paths. On the one hand, chemistry advanced toward organization, with movement from the elementary to the compound, beginning with the way in which particles of atomic material become associated with organized entities. On the other hand, the mechanical organization, in attempting to link with chemistry, has moved from the compound to the elementary, beginning with the entities organized spatially toward the particles of active material of the constituents.

Either explicitly or implicitly, throughout the history of biology, the alternative choice has always been between viewing matter as passive and viewing properties (such as life) as being generated by mechanical or spatial organization, or viewing matter (such as chemicals) as active and incorporating its properties (including life) into the constituent elements. These two alternative paths have also always been present in chemistry, in which the atomic view and the concept of spatial organization in molecules have never been able to explain the novel properties that emerge when a molecule is formed.

On the other hand, the approach of the systemic complexity that we defend tells us that every structured, organized entity or system is different from its constituent parts and its properties are not linearly correlated to those of the constituent parts. This applies to both the system molecule and the system cell. The constituent parts have their own specific properties, and they are not merely passive matter. However, when those parts are joined to form a new system, novel properties emerge. In order to explain the novelty, we stress that the constituents are also themselves systems and that, by virtue of their structure and organization, they are modified when they enter the composition of a new system. The composition of the constituent systems must first be dismantled and then restructured to create the new system with its novel emergent properties. Every chemical reaction is a kind of "death" of old molecular structures and "birth" of new molecular structures.

With this philosophical background in place, let us now discuss the three historical moments in the 19th century during which the opposition between chemistry and biology came to the fore both directly and materially.

Wöhler's Artificial Synthesis of Urea

In 1828, the German physician and chemist Friedrich Wöhler published an article entitled *On the Artificial Production of Urea* in which he described the formation of urea by combining cyanic acid and ammonia. Urea, discovered in 1773 by Hilaire Rouelle as an important component of human urine, was thus artificially obtained for the first time in vitro and so constituted the first synthesis of an organic compound from two inorganic molecules. This experimental work raised both general and more specifically chemical problems. Leaving aside the specifically chemical problems and focusing on the two more general problems, two issues arise. The first issue concerns the relationship between organic and inorganic chemistry, and the second concerns the relationship between chemistry and the substances present in living organisms. In particular, there is the well-known philosophical issue concerning the vitalistic origins of organic substances.

As Peter Ramberg points out, historians today agree that Wöhler's synthesis of urea has had little real impact of its own on the issue of vitalism. It is only with time that this synthesis has been "mythologized." Ramberg claims:

> Equally famous is the mythical story concerning the impact of Wöhler's synthesis on the science of organic chemistry, a story that has been repeated countless times in textbooks, lectures, and articles. The Myth has a number of features. Before Wohler announced his synthesis, the story goes, chemistry was deeply divided into organic and inorganic realms. Organic compounds, derived from plant and animal sources, were less stable, more prone to decomposition, and had compositions more difficult to ascertain by elemental analysis. Whereas inorganic compounds followed the laws of chemistry and were easily analyzed and synthesized, organic compounds could be made only in plants or animals by a mysterious vital force that could not be replicated in the laboratory. Wohler's synthesis of urea from inorganic sources, the mythical story continues, removed this artificial barrier between organic and inorganic chemistry. Chemists then realized that organic and inorganic compounds were governed by the same laws, and Wöhler's synthesis effectively unified chemistry. Furthermore, because Wöhler had succeeded in making an organic compound from its elements, the concept of the vital force was no longer necessary and vitalism could be thrown into the dustbin of failed theories.[1]

---

[1] Ramberg, "The Death of Vitalism and the Birth of Organic Chemistry: Wohler's Urea Synthesis and the Disciplinary Identity of Organic Chemistry."

We believe Ramberg's historical analysis is acceptable, but for our purposes, we should note a few things that Ramberg also partially highlighted. First, Wöhler himself is saying that the term *artificial* in this context means that urea is formed by "man's effort and not by nature." In a famous letter to Berzelius, Wöhler wonders: "Can one consider it as an example of the formation of an organic substance from inorganic substances? It is peculiar that the production of cyanic acid (and also ammonia) always requires an originally organic substance; a *Naturphilosoph* would say that the organic has not yet disappeared from either the animal coal, or in the cyano compounds formed from them, and therefore an organic body can always be reproduced from them."[2]

Here, Wöhler is clarifying his belief that vitalism was a manifestation of *Naturphilosophie*. In practice, he was already considering vitalists' possible objections to his discovery. The fact, then, that other important chemists in the following years positioned themselves on Wöhler's side with respect to vitalism tells us that, though not exclusively, the synthesis of urea was soon considered important in overcoming the view that vitalistic substances are present in living beings. For our purposes, this is the crucial point. In the years following Wöhler's synthesis, vitalism moved from considering living matter as less and less "specific" and different from ordinary matter to considering processes involving living beings. A clear and illustrative example is the importance given to microorganisms in the fermentation process, with the work by the important biologist Louis Pasteur.

The Dispute between Berthelot and Pasteur over Fermentation

In 1857, Pasteur began his work on fermentation, not with an alcoholic ferment but with a particular lactic yeast ferment that he had discovered. He demonstrated that this ferment was (1) a living organism and (2) the active cause of lactic acid production. Pasteur then extended this approach to alcoholic fermentation and reached the conclusion that all fermentative processes required the presence of yeast, a living organism. He concluded that alcoholic fermentation, like other types of fermentation, was a phenomenon related to a vital action. Here are Pasteur's own words on fermentation:

> The works of [Liebig and Berthelot] seek to reject the idea of an influence on the part of organization and of life upon the cause of the phenomena that concern us. I am driven to a completely different point of view. In the first part of

---

[2] "Correspondence from Wohler to Berzelius (February 22, 1828)," 208.

this work, I propose to establish that, just as there exists an alcoholic ferment, brewer's yeast, that is found wherever there is sugar that breaks down into alcohol and carbonic acid, there is also a particular ferment which is a lactic yeast that is present whenever sugar becomes lactic acid and, if any synthetic nitrogenous material can transform sugar into this acid, it is because it is a convenient nutrient for the development of this ferment.[3]

He added, "The conclusions to be deduced from these preceding facts are evident for everyone. The resolution of sugar into alcohol and carbonic acid correlates to a vital phenomenon."[4] In 1860, Berthelot replied:

The facts that have just been exposed throw a new light upon the nature of brewer's yeast and of the phenomena that it determines. In fact, they prove that yeast does not constitute a unique and definite type of ferment. We know that Cagniard's research and especially that of Pasteur have established that brewer's yeast is constituted by a mycodermic plant. Based on the new experiences that I have just reported, I think that this plant does not act upon sugar by virtue of a physiological action but simply by ferments that it is capable of excreting, in the same manner that sprouted barley secretes distase, almonds secrete emulsin, an animal's pancreas secretes pancreatin, and the stomach of the same animal secretes pepsin. Among the ferments that are secreted, those that are soluble may be isolated and purified up to a certain point, in the manner of the immediate principles defined.[5]

Berthelot not only responds that the fermentation of beer that Pasteur examined has no biological origin, but he also lists a whole series of ferments establishing that fermentations are not physiological phenomena (acte physiologique) but rather chemical phenomena. In his 1860 book Berthelot claims:

[I]n all cases of fermentation, we tend to reproduce the same phenomena through chemical methods and to interpret them through exclusively mechanical factors. The goal of our studies is to banish life from all explanations relative to organic chemistry. It is only thus that we will succeed in constituting a science that is complete and independent, which is what it must be in order to effectively contribute to the understanding of physiological changes and to their artificial reproduction.[6]

---

[3] Pasteur, "Mémoire de L. Pasteur sur la fermentation appelée lactique."
[4] Pasteur, "Mémoire de L. Pasteur sur la fermentation alcoolique," 1035.
[5] Berthelot, "Sur la fermentation glucosique du sucre de canne," 983.
[6] Berthelot, Chimie Organique Fondé sur la Synthèse, 655–656.

Here, we note that the chemical approach is not only used to "explain" biological processes but, above all, to reproduce them in the laboratory.

For Pasteur, on the other hand, the biological approach applied not only to fermentation but also more generally. Gerald Geison tells us that

> Pasteur's conception of putrefaction and disease very much resembled his theory of fermentation, as indeed was to be expected from the long-standing and widely accepted analogy between fermentation, putrefaction, and disease. With respect to putrefaction, Pasteur insisted that the "putrefactive ferments per se" were anaerobic microbes belonging to the genus *Vibrionia*. . . . With respect to disease, Pasteur supposed that each infectious disease was linked with the vital activity of a specific microorganism,[7]

Geison adds that "[f]or most of his life Pasteur held a similarly biological conception of the process of immunity. Linking immunity with the nutritional requirements of the pathogen, he suggested that the tissues of the invaded animal might contain only trace amounts of rare substances required for the nutrition of the invading microbe. . . . Pasteur never disavowed his fundamentally biological theory of disease even as he proved willing to adopt a modified chemical theory of *immunity*."[8] Yet "[i]n 1897, within two years of Pasteur's death, Eduard Buchner did finally show that a cell-free extract of yeast juice could produce alcoholic fermentation, and he ascribed this result to a soluble alcoholic ferment or enzyme ('zymase') that he presumed to operate normally within the yeast cell. After a brief period of controversy, Buchner's position was widely adopted. . . . By roughly 1900, then, Pasteur's strictly biological theory of fermentation looked distinctly outmoded."[9]

An important point in the dispute between chemists and biologists regarding fermentation was the conceptual and terminological issue that led to the two terms *ferment* and *enzyme*. The word "enzyme" was invented in 1876 by the University of Heidelberg physiology professor Wilhelm Friedrich Kühne. Berthelot's work of 1860 was highly influential on Kühne's choice of this new word to replace the use of bulky phrases, such as "unorganized" ferment, which until then had been applied to soluble or extractable chemical ferments in order to contrast them with "formed" or "organized" vital ferments.

Kühne also wanted to codify, from a terminological point of view, the differ-. ence between the chemical and the biological approaches. For him, the concept of ferment had to remain inextricably linked to vital processes, while the concept

---

[7] Geison, "Pasteur on Vital versus Chemical Ferments: A Previously Unpublished Paper on the Inversion of Sugar," 428.

[8] Ibid., pp. 428–429.

[9] Ibid., pp. 429–430.

of enzyme was invested with a chemical aura. Both of these terms were attributed to yeast, "ferment" being the old term for yeast and derived directly from the agitation of a sugar solution in fermentation. "Enzyme" literally means "within the leaven" and was used to denote the substance found in yeast. Since the end of the 19th century, the word "enzyme" has completely replaced all other expressions regarding these processes, leading to the disappearance of the vitalistic residues that had been inherent in the word "ferment."

The vitalistic perspective regarding living matter has been completely overcome when there is no scientist or philosopher who considers that there are two different types of substances—those of the living and those of the nonliving—and that the former cannot be identified with the latter because they exhibit recognizable differences, despite similarities between them or even chemical correspondences that are not only functional (physiological) but also physical in the broad sense. Today such an objective has been reached, but, as we shall see, different forms of vitalism have seeped into the more immaterial aspects of living matter.

Before looking at Buchner's work in some detail in order to achieve a closure on the dispute over fermentation, we wish to make a philosophical point about the 19th-century debate between Pasteur and Berthelot regarding the chemical and biological approaches to fermentation. From our 21st-century systemic point of view, we perceive that the opposition of these two approaches cannot exist. One of the fundamental theses defended in this book is that, in fact, even the biological approach to such processes uses a more or less articulated chemical type of explanation. Even if it were true that microbes are needed for fermentation to occur (and we shall see that this is not the case), a complex series of chemical transformations would have to occur within microbes that would constitute both the cause and the explanation for fermentation. And as we shall later see, although the immunological and many other aspects of diseases can be examined with nonchemical approaches, they will also always have a more or less complex reading within the chemical approach.

The Discovery of Buchner's Cell-Free Fermentation

To understand Buchner's work on fermentation and its general importance, we will quote directly from his 1907 Nobel Lecture. He begins by saying that his work is "on the boundary between animate and inanimate nature."[10] He then describes the natural process of fermentation: "If fruit juices or sugar solutions are left to stand in the open air, they show after a few days the processes which

[10] Buchner, "Cell-Free Fermentation," 103.

are covered by the name of fermentation phenomena. Gas is seen to develop, the clear solution becomes cloudy and a deposit appears which is called yeast"[11] and tells us that the role of yeast has "remained for a long time obscure. It was believed that its appearance was of a secondary nature and it was regarded as an inferior kind of precipitation product."[12]

In the 1830s, yeast was finally recognized as composed of the cells of a plant: "Three researchers, Cagnard Latour in Paris, Theodor Schwann in Berlin and Friedrich Kützing in Nordhausen, reported almost simultaneously that yeast consisted of live cells of a plant,"[13] and this vitalistic interpretation was then extended to other fermentations, reaching the conclusion that fermentations are the result of microorganisms. "Kützing extended his investigations not only to yeast but also to mother of vinegar, which converts ethyl alcohol into vinegar. The fermentation processes thus appeared as a result of the life activity of micro-organisms."[14]

This "vitalistic" idea of fermentations was opposed by important chemists, such as Berzelius, Liebig, and Wöhler: "This vitalistic view . . . received a very mixed reception among the naturalists. In particular there was no lack of keen, even derisive criticism from the greatest chemists of the time, Berzelius, Liebig and Wöhler. Berzelius called the new concept of yeast a scientific-poetic fiction."[15] The aversion of chemists is understandable if one examines Buchner's discussion of urea:

> This attitude of total rejection on the part of the foremost chemists is understandable. It was indeed only a few years earlier (1828) that Wöhler had succeeded in artificially producing urea, a substance which had previously been conceived as a type of all substances produced in animal bodies only, under the influence of the life force. No sooner had it been realized, said Liebig, that all life processes in plants, just as in animals, must be conceived as physical and chemical processes, then along came unscientific people and tried to make acts of life out of simple chemical processes.[16]

Buchner then cites Pasteur's experiments, which, according to Berzelius, settled the dispute between chemists and biologists regarding whether or not fermentation required yeast: "It was only the systematic and striking experiments of Louis Pasteur, extending over a decade, which finally led to the recognition that

[11] Ibid.
[12] Ibid.
[13] Ibid., 104.
[14] Ibid.
[15] Ibid.
[16] Ibid., 105.

in Nature without living organisms, without live yeast, no fermentation exists. This put an end to all disputes. Fermentation was seen to be a physiological act inseparably linked with the life processes of yeast."[17] Buchner adds that Moritz Traube had already established, in 1858, that fermentation was caused by chemical substances in microorganisms called enzymes:

> A simpler assumption was made by Moritz Traube in Berlin (1858), according to which there was in micro-organisms a certain chemical body which caused fermentation. Similar substances, chemically very active, which are known today as enzymes [but] although enzyme theory has found many supporters, so far it has not been possible to separate the enzyme from yeast cells. "This enzyme theory as an explanation of fermentation phenomena found great favour in wide circles . . . every attempt made to separate such an enzyme from the yeast cells had failed.[18]

Buchner then describes in detail his experiment, how he eliminated yeast cells, and how he produced a pressed yeast juice that still produces fermentation:

> Pressed yeast juice, a yellow-brown liquid, of which I am here handing round samples, smells pleasantly of yeast and in transmitted light appears transparently clear, though in incident light it appears opalescent. On heating, it soon separates flakes of coagulated protein and on further heating this formation may be so extensive that when the container is inverted scarcely any liquid flows out. . . . If sugar solution is added to freshly expressed yeast juice, a strong formation of gas sets in after a little while. The active frothing of carbon dioxide bubbles and the formation of a thick layer of foam show that a fermentation process has started.[19]

Buchner, therefore, makes a first general point: "The presence of coagulable protein in the interior of micro-organisms has thus been established for the first time."[20] Buchner proceeds to responds to some objections concerning the elimination of yeast cells from the sugar solution in his experiment:

> The first question now was whether the few yeast cells still present in the expressed juice could in any way be considered to be the cause of the decomposition of the sugar. The answer to this is certainly in the negative, since their number is much too small. The juice can be filtered through a diatomite filter

---

[17] Ibid., 106.
[18] Ibid., 107.
[19] Ibid., 111.
[20] Ibid.

and even through biscuit-porcelain candles, without its action being completely destroyed.... Furthermore, there was reason to assume that the fermentation action of expressed juice was attributable to pieces of live plasma present. This assumption contradicts in particular the behavior of the expressed juice when antiseptic media are added. Especially, it is seen that toluene actually does prevent the action of live yeast on sugar, though not the action of expressed juice.[21]

Next, he cites a series of further experiments that he performed to show that no yeast cells are involved and that it is the enzyme that produces the fermentation:

A whole series of further experiments conclusively decides against that hypothesis. First of all, the expressed juice can be concentrated by evaporation at low temperature in a vacuum chamber and finally fully dried. The yellowish residue, reminding one of dried egg yolk ..., is mainly soluble in water and still shows unchanged the fermentation action on sugar. Further, if the expressed juice is added to a quantity of alcohol and ether, a white precipitate is obtained, which can be transformed in the vacuum chamber into a dust-dry powder.... This also is largely soluble in water and causes strong fermentation when sugar is added. Finally, the yeast cells can be killed, without their fermentation action being destroyed.... one can add the live yeast cells to large quantities of alcohol or acetone and finally wash them with ether. The air-dried "permanent yeast" thus obtained, also called "zymin," is incapable of growth but, when sugar solution is added, can produce an extraordinarily powerful fermentation.[22]

Buchner reaches the following conclusion regarding cell-free fermentation by a chemical: "If we now summarize the results of all these experiments, we establish that a separation of the fermentation effect from the live yeast cells can be carried out. To start off a fermentation process, no such complicated apparatus is needed, as the yeast cell is; rather is there a 'cell-free fermentation.' ... The active agent in the expressed yeast juice appears rather to be a chemical substance, an enzyme, which I have called 'zymase.' From now on one can experiment with this just as with other chemicals."[23]

For Buchner, the dispute between chemists and biologists over fermentation was not settled in favor of either side: "The differences between the vitalistic view and the enzyme theory have been reconciled. Neither the physiologists nor the chemists can be considered the victors; nobody is ultimately the loser; for the

---

[21] Ibid., 112–113.
[22] Ibid., 113.
[23] Ibid., 114.

views expressed in both directions of research have fully justified elements. The difference between enzymes and micro-organisms is clearly revealed when the latter are represented as the producers of the former, which we must conceive as complicated but inanimate chemical substances."[24]

According to Buchner, we can extend the example of fermentation, just as we extended the example of urea:

> In one case, which seemed typical of processes strongly linked with the whole life of the cells, it was possible to trace the whole phenomenon to the relatively simple action of a chemical body in the interior of the cell. Just as it was earlier learnt how to produce urea without a living animal, in the test tube, without any life force, so it is seen here that an apparent action of live cells can take place without cells" and arrive at the general conclusion that cells are "chemical factories": We are seeing the cells of plants and animals more and more clearly as chemical factories, where the various products are manufactured in separate workshops. The enzymes act as the overseers.[25]

It seems clear just how modern this conclusion is and how much it relates to the general approach that is taken in this book.

Energy/Heat in Living and Physiological Chemistry

In the early part of the 19th century, the study of substances in isolation from living systems was separate from general chemistry and was linked in a clear way to physiology. Beginning with the 1840s, a distinction was drawn between the scientific study of the elementary composition of substances and investigation of their physiological role. In this way, two new disciplines were born that clearly divided experimental objects into distinct categories.

On the one hand was organic chemistry, which arose at the end of a long theoretical process and was defined as the chemical study of carbon compounds in isolation from vital systems. On the other hand was physiological chemistry, which devoted itself instead to the behavior of organic substances within living organisms. The study of chemical transformations within the organism had proved too difficult and complex to be treated in strictly chemical terms and had been outside the theoretical and experimental reach of 19th-century chemistry. Justus von Liebig played a pivotal role in the disciplinary transformation of this field of research. Although he was bound to the earlier research program, he

---

[24] Ibid., 119.
[25] Ibid.

initiated a school and an experimental practice that led the direction of research over the following decades.

From 1840 to 1841, Liebig devoted himself with great interest to the application of organic chemistry to physiology and pathology, although, in his training, he had no real focus on the typically medical and physiological aspects of chemistry. For Liebig, physiological chemistry could only provide quantitative data on the composition of substances in order to evaluate input and output balances and to locate the material conditions of basic physiological processes.

For Liebig, the role of chemistry in physiology was limited to the more precise characterization of phenomena and to control of the precision of observations by means of "number and weight." Liebig described the animals as "walking stoves" governed only by what is put in and the heat that is produced. In contrast, chemists such as Karl Gotthelf Lehmann and Claude Bernard, working with a more physiological focus or physiologists in the strict sense, insisted on research within the animal organism. They did not simply intend to observe and measure what was entering and leaving such an organism. For Liebig, the body was a sort of black box whose interior was not discoverable by the theoretical and experimental tools available to the chemist and which, therefore, could only be studied from an operational point of view.

As a result of this work, Liebig published a book on plant chemistry in 1841 and in the following year, a book on animal chemistry (*Thierchemie*), which played a central role in the subsequent decades. The latter book is dedicated to the exposition of classes of organic substances involved in the main physiological functions. Following William Prout's proposed classification system, these classes are distinguished as either sugars, fats, or nitrogenous substances, and each is assigned a specific role within the economy of the animal organism. Liebig began to support Johann Friedrich Gmelin's theory that the fundamental constituents of animals were synthesized in plants and then gradually oxidized in animals.

Based on this theoretical position, Liebig proposed to deduce the main chemical and physiological pathways underlying bodily functions. The development of knowledge regarding the elementary composition of many organic compounds, to which the Liebig's group was making important contributions, also allowed scientists to hypothesize the possible chemical changes and the different compounds into which the elements are transformed during their passage inside the body. With the comparison between the elemental composition of food, tissue constituents, excreta, and respiratory gas, one could expect transformations of these constituents (which Liebig called "metamorphoses") as well as determine the anatomical sites in which these changes could occur.

Liebig's work summarized and systematized all previous chemical knowledge of physiological phenomena and helped to establish new standards of scientific rigor. Even the hypothetical equations, which were the result of the widespread attitude toward the mathematization of his time, would indicate to future researchers that organic compounds were formed through chemical and physiological processes that could be described by chemical reactions that were yet to be discovered. This indication would help to build the research program first of physiological chemistry and then of biochemistry.

A new physiological chemical program emerged in the 1870s within the German mechanistic physiological tradition. The aim of this program was to clarify the origin and location of life forces and the transformation of energy from synthesis to animal heat, metabolism, nutrition, and the nature of muscle action. Eduard Buchner's discovery, which we have already discussed, shifted the early 20th-century emphasis on enzyme mechanisms, and the purpose thus shifted toward discovery of the different enzymes responsible for chemical functions within the cell. In this period, the cell was thought to be a "chemical factory" in which many different chemical tools worked and could be analyzed. In a famous 1901 essay on the biochemical organization of the cell, Franz Hofmeister stated that, in the cell, the synthesis and degradation of various substances proceeded by means of a series of intermediate steps that did not always include the same type of chemical reactions. Rather, they included a series of reactions of different types. These series of reactions needed to be studied and would then form the metabolic pathways and cycles that we will discuss later in this book.

In the second half of the 19th century, Claude Bernard postulated one of the fundamental concepts of modern biology, the "internal milieu." The constancy of the internal environment was the condition for liberating a living organism from the "external space," that is, from the external environment in which this organism lived. As Bernard explained:

> The living body, although it needs a surrounding environment, is nevertheless relatively independent of it. . . . The constancy of the internal environment is the condition for free and independent life: the mechanism that makes this possible is the one that ensures the maintenance, in the internal environment, of all the necessary conditions for life. . . . The constancy of the environment presupposes a perfection of the organism such that the external variations are at every moment compensated and kept in equilibrium. Consequently, far from being indifferent to the outside world, the superior animals are on the contrary in close relationship with it, so that their equilibrium results from a continuous and delicate compensation with a very sensitive balance.[26]

---

[26] Bernard, as cited in: Pennazio, "Homeostasis: A History of Biology."

The development of biochemistry began between the end of the 19th and the beginning of the 20th centuries. Two sciences that contributed strongly to this were microbiology, which is concerned with the discovery of toxins and antitoxins and with considering infections as the result of biochemical processes, and immunology, which elaborates chemical models for the functioning of antibodies. Furthermore, the development of the concept of the hormone as a "chemical messenger" provided new space for developing theories about the chemical ordering of the organism. However, the first powerful formulation of this concept of chemical ordering could already be found in the concept of a chemically regulated "internal environment."

Order/Disorder and the Complexity of Living Beings: Finalism and the Evolution of Species in Darwin, Spencer, and Bergson

Two important advancements of 19th-century science were the birth of modern biology and, in a more strictly physical context, the birth of thermodynamics. As we shall see, these two 19th-century scientific branches were not easily reconcilable and would not be fully integrated until the 20th century. We shall see in a following chapter that the Second Law of Thermodynamics was eventually reconciled, leading to "a destruction of order and an increase of disorder" within the organic realm. However, this integration is not even considered complete today, and it is not uncommon to find philosophical discussions that begin with the assumption that life either is in opposition to the laws of thermodynamics or is at least poorly explained in thermodynamic terms. In that later chapter, which concerns the 19th-century relationship between biology and thermodynamics, we will be concerned with two aspects that will converge into a critique of Darwinian evolutionism: one linked to the biological individual and the other to the species.

It is not our purpose to delve deeply into Charles Darwin's enormous influence on modern biology. In this regard, we simply wish to emphasize the one important factor against which Darwinian evolutionary theory has always had to struggle, that is, the apparent tendency toward order that is evident in the evolution of species. In fact, however, the fundamental truth that Darwin highlighted was twofold: First, species have changed over time, and, second, species have changed by virtue of a general phenomenon that he called natural selection. This, however, did not imply any sort of teleology or finalism. From the very beginning, Darwin sought to demonstrate that a "transmutation of species" had occurred, transmutation being a term that is much more appropriate to his theory than the term *evolution*. He later made free use of the term *transformation*, in the sense of "change of form," and he defined his view as a theory of modification (of the species) by natural selection. Darwinian theory employed the expressions change, variation, transformation, transmutation, and mutability to connote the evolutionary hierarchy, along with chain of being, tree of life, and organization of life.

Natural selection is considered the cause of the origin of one species from another. Therefore, according to Darwin, the theory of the origin of species is inconceivable without the theory of natural selection. Since the origin is the first moment of the generation of species, natural selection is a necessary condition of this process. Over time, Darwin concluded his doctrine by adding sexual selection as a vehicle of species transformation and by explaining the generation of human beings. He argued that, to a small but important extent, habit and adaptation to the environment also helped explain the origin of species. But these processes never took the place of the important role of variation by natural selection.

It is erroneous to reproach Darwin by accusing him of imagining natural selection as a choice made by nature. On the contrary, he imagined a nature within which everything occurred as if choices had been made, as if there were a teleology governing the process, although there is no such teleology or choice. That is, although there is no ordering purpose that drives the process of transformation of species, the end result appears orderly. Thus, the actual process of transformation described by Darwin in no way contradicts either thermodynamics or the tendency toward entropy. Through the play of natural forces alone, such as the tendency toward spontaneous variation, the struggle for life determined by the scarcity of means of subsistence, and the elimination of the less adapted members of a species, a species becomes adapted to its environment. Therefore, there is a transformation of the old iteration of a species and an increasingly satisfactory adaptation of the new iteration of that species to its conditions of existence, without needing to hypothesize a particular type of teleological causality whose task is to direct this process of transformation.

Although this approach could also be used to explain the structural order and perfection of individual organisms, such an explanation never succeeded in convincing all biologists. The reason is clearly discussed by Richard Dawkins in the first chapter of his work, *The Blind Watchmaker*. Dawkins states:

> The watchmaker of my title is borrowed from a famous treatise by the eighteenth-century theologian William Paley. His *Natural Theology – or Evidences of the Existence and Attributes of the Deity Collected from the Appearances of Nature*, published in 1802, is the best-known exposition of the "Argument from Design", always the most influential of the arguments for the existence of a God. It is a book that I greatly admire, for in his own time its author succeeded in doing what I am struggling to do now. He had a point to make, he passionately

believed in it, and he spared no effort to ram it home clearly. He had a proper reverence for the complexity of the living world, and he saw that it demands a very special kind of explanation. The only thing he got wrong—admittedly quite a big thing!—was the explanation itself. He gave the traditional religious answer to the riddle, but he articulated it more clearly and convincingly than anybody had before. The true explanation is utterly different, and it had to wait for one of the most revolutionary thinkers of all time, Charles Darwin. Paley begins *Natural Theology* with a famous passage: "In crossing a heath, suppose I pitched my foot against a stone, and were asked how the stone came to be there; I might possibly answer, that, for anything I knew to the contrary, it had lain there forever: nor would it perhaps be very easy to show the absurdity of this answer. But suppose I had found a *watch* upon the ground, and it should be inquired how the watch happened to be in that place; I should hardly think of the answer which I had before given, that for anything I knew, the watch might have always been there."[27]

On the other hand, Jacques Monod focuses on differentiating random and necessary aspects of biological evolution and states: "For modern theory *evolution is not a property of living beings*, since it stems from the very *imperfections* of the conservative mechanism."[28] And then Monod adds:

The initial elementary events which open the way to evolution in the intensely conservative systems called living beings are microscopic, fortuitous, and utterly without relation to whatever may be their effects upon teleonomic functioning. But once incorporated in the DNA structure, the accident—essentially unpredictable because always singular—will be mechanically and faithfully replicated and translated: that is to say, both multiplied and transposed into millions or billions of copies. Drawn out of the realm of pure chance, the accident enters into that of necessity, of the most implacable certainties.[29]

At first glance, the title of Darwin's main work is very explicit on this issue. Since the first edition in 1859, the title of his work has never changed: *On the Origin of Species by Means of Natural Selection, or the Preservation of Favoured Races in the Struggle for Life.* From the very beginning, there was a serious inaccuracy in the definition of the topic of the book itself. This is because Darwin does not clarify anywhere what he means by "the origin of the species" or whether he is referring to the very existence of a species itself. He does not ask how the various

[27] Dawkins, *The Blind Watchmaker: Why the Evidence of Evolution Reveals a Universe without Design*, 7–8.
[28] Monod, *Chance and Necessity. An Essay on the Natural Philosophy of Modern Biology*, 116–118.
[29] Ibid.

species might have originated; rather, he is simply concerned with the fact that the species exist and with explaining how they have come to be as adapted as they are. The problem of the absolute point of origin of species is nowhere treated by Darwin, although he may only incidentally allude to it. Assuming that we agree on the meaning of the term *origin*, the term *species* still remains to be defined. Everyone has a general idea of what it means: A species is a group of plants or animals that share similar traits and that can be easily distinguished from other groups. No one hesitates to distinguish an individual of the swallow species from an individual of the elephant species.

The difficulty occurs when, in considering any species, one attempts to describe the characteristics that define it, what philosophers call the "necessary and sufficient conditions" for belonging to that particular species. This is problematic because such conditions are impossible to identify since, within a specific species, one will not find two identical individuals. One will merely find what the philosopher Ludwig Wittgenstein called family resemblances among such individuals. Considering only those individuals who are fairly similar to each other and can, undoubtedly, be classified in the same group, one will soon realize that there are subgroups or subspecies to be considered, as well as varieties that initially appear to be within that species but then may also be found in other species. Even Darwin stumbled upon this problem, initially dividing a species into varieties, then bringing it back to the unity of the species, doing and undoing the same work twenty times but failing to find a reliable criterion to define membership in a given species.

With the exception of Georges-Louis Leclerc (a.k.a., the Comte de Buffon) and Jean-Baptiste Lamarck, all of Darwin's predecessors believed that existing species had always existed and regarded them as being fixed. Their position seems, at first, to be the most coherent since it is not easy to define a species other than as a class of living beings with characteristics irreducible to those of any other class. Therefore, a species is by definition a rigidly defined typology for which changing would mean ceasing to be itself and to exist. However, to say that species are fixed is tautological because to suggest that they change is tantamount to saying that they do not exist. Why would Darwin insist on saying that species change rather than simply saying that they come into being and then become extinguished?

Two species are considered truly distinct when their mating is sterile. Darwin is careful not to deny such evidence, but he questions it. You can see that this fact worries him a little. According to this theory, the fertility that is present when varieties originating from a common ancestor mate, or even the fertility of their mixed offspring, is as important as the sterility of a species because this is what broadly and clearly distinguishes between species and varieties. The problem was and still remains true: "neither sterility nor fecundity provides a clear distinction between species and varieties; the proofs obtained following this criterion are as uncertain and doubtful as those drawn from other structural and constitutional differences."[30] Paradoxically, the theory of the transformation of species establishes, first of all, that their genetic fixity is the most evident sign of their identity as a species.

This paradox, however, is only apparent. One must always keep in mind that all entities and their properties are inherently dynamic, although this view was taken to the extreme by Heraclitus, whose philosophy embraced only dynamism and change and thus led to the rejection of any stable entities. A set of properties that seems static to our senses and instruments defines an entity, but, in fact, these properties and the entire entity are slowly changing for the "measuring instrument" used to observe them. There is nothing really static about the universe. The universe itself has a history, which makes it impossible to set a time whose validity is no more than conventional. However, not everything changes at the same speed, and it is the different rate of transformation of the various properties that allows us to talk about "entities" as static objects and as independent of time.

The different rates of transformation determine the scale of time frames that, given the scales of size and energies, allows us to reduce the evolving complex into a simple and stable entity, made of a static aspect and a dynamic transformation. It is, in fact, the scale of the time frames that allows us to conceptually differentiate the various processes at play, thus canceling the processes that operate slowly in the interval of the time frames under examination and obtaining properties and entities that are undergoing real transformation but are static with respect to the processes involving properties. Such separation is never, however, absolute.

This applies to both living beings and inanimate matter, as well as to concepts. In the transformation of biological species, nature has been magnanimous. The time scale of these changes is enormous. It is, in fact, practically infinite compared to the lifespan of individuals. This allows us to define a stable species within the time frames of individuals and to ignore their individuality within the time frames of species. After all, this situation is more simple than the chemical one in which the species' scale of existence and the timing of some dynamic properties can often be similar. It is this situation that gives the concepts of entities (atom and molecule on the microscopic level and element and compound on the macroscopic level) and their transformability (reactivity) equal importance in chemistry. Nevertheless, the concept of the chemical species is fundamental; therefore, there is no reason for these biological concepts to be considered problematic.

---

[30] Kölreutere and von Gärtner, as cited in Étienne Gilson, *Biofilosofia Da Aristotele a Darwin e ritorno*, 233.

Herbert Spencer is responsible for making the notion of evolution the key concept of the second half of the 19th century. The fusion of Darwinism and Spencerism was almost instantaneous, in spite of Darwin's and Spencer's protests. As Spencer tells us:

> Most people admit with no hesitation that Darwin's doctrine, the hypothesis of natural selection, and that of organic evolution are one and only thing. Yet there is a difference between them analogous to that which separates the doctrine of gravitation from that of the solar system; and as the latter, admitted at the time of Newton, could have been sustained even if Newton's law had been rejected, in the same way the rejection of natural selection would leave unchanged the hypothesis of organic evolution.[31]

Claiming the authorship of the theory of evolution in general, and of organic evolution in particular, Spencer postulates "adaptation to conditions" as a general cause of transformation. In short, with regard to the precise issue of cause and of how evolution occurs, Spencer might have been called a Darwinian, but he is, rather, more of a Lamarckian. In Darwin's autobiography, he tells us:

> I don't think the knowledge of Spencer's works had any influence on my work. The deductive method by which he deals with every subject is completely contrary to my mental approach. His conclusions never convinced me, and I never failed to repeat myself, after reading his discussion: Here is a topic that would require six years of work! Its fundamental generalizations, the importance of which has been compared by some to that of the laws of Newton (!) and of which I also admit that they are of great value from the philosophical point of view, do not seem to me of any scientific utility. They have more the character of definitions than of laws. They do not help to predict what will happen in any particular case. At least, they were not useful to me.[32]

Far from being tangled and unclear, Spencer's concept of evolution is a prodigious generation system in which each moment adds something new to the previous one. Thus, we are already dealing here with the notion of creative or, at least, progressive evolution.

Henri Bergson, for his part, fuses the two concepts of evolution and progress. But progressivism is not a necessary component of the idea of evolution. Even if we can admit progress at the level of the whole, we must also recognize regress as occurring in the parts. Neither Buffon, who was sensitive above all to the

---

[31] Spencer, as cited in Étienne Gilson, *Biofilosofia Da Aristotele a Darwin e ritorno*, 102.
[32] Darwin, as cited in Étienne Gilson, *Biofilosofia Da Aristotele a Darwin e ritorno*, 111.

"degeneration" of the species, nor Darwin, who consoled himself at the thought that death is usually quick and painless, would allow themselves to embrace an excessive confidence in a bright and progressive future. By contrast, Spencer's early writings are titled *Essays on Progress*, and Bergson titled his main work *Creative Evolution*, never questioning that the universe is constantly progressing and that we must even accept the death of individuals for the sake of a greater progress of life in general.

If the theory of evolution is correct, this means that all processes and events occur over time. This is also why evolutionary biologists have always been concerned about there being sufficient time for this evolution to take place. At this point, we must insert the contribution of thermodynamics. In 1862, the physicist William Thomson, a.k.a. Lord Kelvin, published his thermodynamic calculation of the age of the Earth as being between 20 and 400 million years old. He later revised his calculation by assigning the Earth an age of 98 million years. However, this age collides strongly with the age of the Earth as determined by geology. Today, we know that the age of the planet Earth is much greater than that calculated by Thomson, as the Earth is approximately 4.5 billion years old. The age attributed to the existence of life on our planet has been extended to about 4 billion years. Thus, Lord Kelvin's thermodynamic calculations have been largely rejected. However, his thermodynamic calculations, which were based on the physical-mathematical approach, were very influential in the second half of the 19th century. They were considered to be much more reliable than the biological or geological calculations. and the fact that the age of the Earth, as calculated by Lord Kelvin is rather short has always been used to question Darwinism itself.

10

The Quantum Revolution

Before introducing the specificities of quantum mechanics, those referred to as "revolution" in the title of this chapter, we will analyze the essential aspects of classical mechanics. Here we consider a physical entity that is isolated from the environment and that mutates over time. In physics such an entity is described by a tuple (i.e., a finite list of elements) of observable physical quantities. The *tuple* can be divided into two main groups: the M quantities that are constant (or approximately constant) in the range that interests us and the N-M remaining physical quantities that change substantially. The former, called system variables, define and describe the system under consideration; the latter are the "configuration variables" (or state variables) of the system and instantaneously describe its state. This division in a system and its state is linked to the interval under examination (a priori all the variables change) and to the precision of the measurement, that, is to the instrumentation used to measure the variables. The number of state variables (in our case N-M) gives us the number of "degrees of freedom" of the system, a number that is a uniquely fixed feature for a given system in mechanics.

The physical quantities that must represent such degrees of freedom are not determined. Such quantities are called "generalized coordinates" of a suitable space or "configurations." This space is essentially static and can therefore only be used to describe the equilibrium states of the system. If we also define the "generalized velocities" (or the "generalized moments"), that is, the derivatives with respect to the time of the generalized coordinates (they will also be N-M), we can define a "phase space" (of dimension 2(N-M)) where a point in that space identifies the dynamic state of the system.

In the 19th century, William Rowan Hamilton formulated the dynamic problem in this abstract way: no longer in terms of spatial coordinates and their derivatives with respect to time (velocities), but introducing pairs of variables p and q (generalized coordinates and moments) considered as independent variables and linked by a fundamental relationship that defines them as canonical variables. Thus, the same system can be described by several pairs of different canonical variables. In this formalism, the energy of the system (the sum of kinetic and potential energy) plays a fundamental role. Expressed in canonical variables, it constitutes the Hamiltonian system, which is left unchanged by all canonical transformations and by all possible changes of variables. The

From the Atom to Living Systems. Marina Paola Banchetti-Robino and Giovanni Villani, Oxford University Press.
© Oxford University Press 2023. DOI: 10.1093/oso/9780197598900.003.0011

Hamiltonian system allows us to express the equations of evolution of the coordinates and generalized velocities in the form of 2(N-M) differential equations of the first order.

In mechanics, there are only two forms of energy: kinetic and potential. Kinetic energy is linked to generalized velocities and is always expressed in a quadratic form. Potential energy, however, is linked to generalized coordinates and can take several mathematical forms in different systems. The sum of kinetic energy and potential energy is the total energy of the system and must be conserved. An important point to highlight is that in mechanics there is a substantial difference between kinetic and potential energy. While the first may always be written as a sum of contributions, each pertaining to a single degree of freedom, it is not always true that this also applies to potential energy.

If it were always true that both forms of energies can be written as the sum of N-M contributions, each relating to a single degree of freedom, there would be N-M independent dynamic equations, one for each degree of freedom. For each, an energy conservation principle would be valid, and no exchange would be possible between the different degrees of freedom that would describe a noninteracting N-M-degree system. Conversely, a potential energy that is an undecomposable function of the coordinates will constitute a "common reservoir" through which the different degrees of freedom can, as a final result, exchange kinetic energy.

Every real system in classical mechanics is thought to consist of a set of "material points," that is, points with mass, each of which has three degrees of freedom represented by its spatial coordinates. In mechanics, therefore, each physical system consists of a set of P particles (atoms, molecules, etc.) with 3P degrees of freedom, and its state at each instant is determined by 6P values. In mechanics there are general laws of motion that translate into a set of differential equations. Once solved, these equations allow us to obtain the functions that describe the temporal evolution of all the material points. These functions give us a trajectory in phase space for each individual material point, and the set of trajectories gives us the transformation in time of a state of the system. Each trajectory remains entirely fixed once one of its points is known, that is, determined by the initial conditions. This trajectory then describes the complete history (past, present, and future) of our material point; the whole of the material points describe the complete history of the system, which is entirely determined and deterministic.

Thus, we can summarize the mechanical scheme as follows: each body can be decomposed into a set of particles, each with its own kinetic energy and with a potential energy due to the global configuration of the whole. For each of these particles, a trajectory in the space of the phases uniquely determines its temporal evolution, and past and future have the same determined value. As Jean le Rond d'Alembert stated 250 years ago, in mechanics time appears as a simple "geometric parameter." From the scientific point of view, this is because the equations of the mechanics are invariant with respect to the temporal inversion $t \rightarrow -t$. As Lord Kelvin (William Thomson) explained: "In abstract dynamics, an instantaneous reversal of the movement of each particle in motion in a system is the cause of a motion backward of the system itself, with each particle along its old trajectory so as to have again, for each position, the same speed in absolute value already assumed before. This is equivalent to saying, in mathematical language, that each solution remains a solution when t is replaced by $-t$."[1] As Henri Bergson pointed out in *L'Évolution créatrice* (1907), in classical physics everything is given, change is nothing more than the refusal of becoming, and time is only a parameter that is not influenced by the transformation described by it.

Subatomic Particles and Atomic Structure

As we have seen, the concept of the atom that was rediscovered at the birth of modern science had largely adopted classical Greek atomism. Pierre Gassendi had eliminated some of the general difficulties that opposed the classical atomic vision to Christianity by eliminating the halo of atheist materialism that had surrounded atomic corpuscles. However, the atomic vision suffered from other general problems. Atoms were theoretical and metaphysical entities that could not be observed, touched, or manipulated by the senses. Certainly, John Dalton attributed to chemical atoms, in a transposed way, a weight that had made them a little more physical, but he never provided any proof of their existence. Many of Dalton's contemporary chemists made a clear distinction between this atomic vision and the experimental evidence that the elements combined in definite proportions. In practice, many chemists adopted a conventional system of (atomic) weights for each element, and the atomic vision was the logical basis with which to assign these weights. But, as occurred for Antoine Lavoisier in the definition of element, many chemists did not pose the problem of the reality of atoms.

By 1837, the confusion was so great that Jean-Baptiste Dumas famously stated that, if it were possible, he would eliminate the term *atom* from science. The main reason behind Dumas and other chemists' position on this issue was the rejection of any speculation that was too far removed from the experimental facts. Chemists needed to see with their eyes and not with their minds. They had to develop theories about facts and not search for facts in order to support preconceived theories. Additionally, many chemists saw atomic theory as a sterile

---

[1] Thomson, (Lord Kelvin), *The Kinetic Theory of the Dissipation Energy*, in *I nomi del tempo*, 197.

paradigm that also led to confusion in chemistry. From a philosophical point of view, they had a preference for Auguste Comte and his positivism.

By the early 20th century, the situation had changed substantially, and atomic theory was virtually universally accepted. If it makes sense in the history of science to use a date as a watershed, the publication of Jean Perrin's 1913 book *Les Atomes* marked the definitive victory of atomism. One example that explains this victory is that, by using thirteen different experimental methods, Perrin determined the value of the Avogadro number, and all were consistent, showing unequivocally that atoms could be "counted." Perrin's success also convinced Wilhelm Ostwald and Henri Poincaré, who had been skeptical of the actual existence of atoms, to change their minds. Poincaré states: "the atomic hypothesis has recently acquired enough credence to cease being a mere hypothesis. Atoms are no longer just a useful fiction; we can rightfully claim to see them, since we can actually count them."[2]

Additionally, two discoveries helped to change the fate of atomism: the discovery of radioactivity and that of the electron. At the end of the 19th century, Henry Becquerel and Pierre and Marie Curie discovered that in nature the transmutation of some chemical elements is spontaneously realized: all the radioactive elements, emitting either alpha, beta, or gamma rays, decayed and formed other elements. For the first time, in 1934, Irène Joliot-Curie (daughter of Pierre and Marie Curie) and her husband Frederic Joliot produced artificial radioactive elements. In 1940, Edwin Mattison McMillan and Philip Hauge Abelson obtained a chemical element with atomic number 93 that did not exist in nature. and they named it Neptunium.

The discovery of radioactivity revolutionized two of the main concepts of chemistry: the element and the atom. The 20th century gave us a perspective on these two concepts that was different from those of Lavoisier and Dalton, who had founded modern chemistry on the basis of these two notions. We shall return later to the 20th-century conception of atom. At this point, however, we wish to emphasize that even the concept of chemical element no longer conforms to what was envisioned by Lavoisier. A chemical element is no longer conceived as a nonconvertible unit of matter because radioactivity transforms one chemical element into another. However, this scientific truth only partially changes the chemical optics. For chemistry and chemists, the chemical element continues to have an almost 19th-century meaning even today, although chemists know that the transmutation of the elements, which was the dream of alchemists, may indeed be a possibility.

Joseph John (J. J.) Thomson's "discovery" of the electron in 1897 changed the perspective in the study of the atomic world and introduced the concept of atomic

[2] Poincaré, in *The Atom in the History of Human Thought*, 256.

structure. Later in this chapter, we will discuss its influence on what "holds together" atoms in their aggregation and, therefore, on the concept of chemical bond. Here, we wish to focus on the birth of the concept of atomic structure. This structuring of the atom has considerable consequences, from both a scientific and a philosophical point of view. Returning to Thomson, he discovered that the cathode rays produced by an electric discharge in a tube containing rarefied gases were formed by particles carrying a negative electric charge. This proved the corpuscular nature of electricity. Once Robert Millikan determined the electric charge of this particle, for which George Johnstone Stoney coined the name electron, the ratio between its mass and its electric charge was determined to be 1836 times smaller than that of the hydrogen ion.

Thomson later demonstrated that electron properties were the same regardless of the type of gas that generated the cathode rays. From this basis, he deduced that electrons were the constituents of all atoms "These particles are called electrons or corpuscles, and no matter what the nature of the gas may be, whether it is hydrogen, helium, or mercury vapour, the electrons or corpuscles remain unchanged in quality; in fact, there is only one kind of electron, and we can get it out of every kind of matter. The conclusion is irresistible that the electron or corpuscle is a constituent of every atom, and that we are able, by forces which we have even now at our command, to detach it from the atom."[3] Additionally, in his Nobel Lecture of 1906, Thomson stated: "The corpuscle appears to form a part of all kinds of matter under the most diverse conditions; it seems natural therefore to regard it as one of the bricks of which atoms are built up."[4]

The presence of electrons in atoms led to many questions, including the following three main ones: how many electrons are present in each atom, where does the positive charge reside, and how do we explain the mass of atoms? These problems were the ingredients of a single concept: the atomic structure. Practically at the same time, Lord Kelvin and John Joseph Thomson proposed the first atomic structure. Kelvin's 1902 model provided for a uniformly distributed positive charge cloud in which electrons were inserted to achieve a situation of balance and neutrality. In 1904, Thomson proposed an atomic model with electrons inside a sphere of positive charge. This model was then called "plum-pudding," as though the electron was incorporated as "plums" in a "pudding," although this image was substantially different from Thomson's original proposal. In 1911, Ernest Rutherford elaborated the planetary model of the atom. The central element of this atom was the nucleus, much smaller than the atom, with a positive charge and carrying almost the totality of the entire mass. The electrons revolved around it, and each type of atom was characterized by its

[3] Thomson, *The Atomic Theory*, 11.
[4] Thomson, *Carriers of Negative Electricity: Nobel Lecture (December 11, 1906)*, 149.

number of electrons and then, to account for neutrality, by its nuclear charge. The Rutherford atom was an atom in which emptiness predominated. In 1920, Rutherford gave this atomic nucleus the name of "proton."

Within atoms, the gravitational force that holds the Sun and planets together had been replaced by Coulomb's force of attraction between the opposite charges of the nucleus and electrons. However, following the pattern of classical electrodynamics, an electron that moved under these conditions would continuously lose energy by emitting electromagnetic radiation and, with a spiral motion, end up on the core. Rutherford's model, therefore, gave a satisfactory explanation of the experiments of Hans Geiger and Ernst Marsden, but it did not explain the stability of the atomic system. A further problem was to explain, using this model, why atoms in emission spectra emitted energy only at specific frequencies, while the Rutherford model predicted a continuous variation in frequency.

A possible answer to these two questions was given by Niels Bohr's 1913 atomic model. Bohr centered his attention on understanding the stability of matter. He stated:

My starting point was not at all the idea that an atom is a small-scale planetary systems and as such governed by the laws of the astronomy. I never took things as literally as that. My starting point was rather the stability of matter, a pure miracle when considered from the standpoint of classical physics. By "stability" I mean that the same substances always have the same properties.... This cannot be explained by the principles of the classical mechanics, certainly not if the atom resembles a planetary system.[5]

Bohr made a first hypothesis that, contrary to the theory of classical mechanics, electrons could only move in certain selected orbits with constant energy or, equivalently, at certain fixed distances from the nucleus (circular orbits). This hypothesis led to an integer and positive number (later called the *principal quantum number*) that characterized the energy of these orbits. A second hypothesis established that all radiative processes in an atom, such as the emission or absorption of photons, were associated with transitions or "jumps" between two orbits (stationary states). These jumps explained the discontinuous variation in atomic absorption or emission spectra, but they immediately opened up another problem: was it such an instantaneous transition? If not, where was the electron during the jump? This concept of transition between states was criticized by many physicists.

---

[5] Bohr, speaking at a conference in Göttingen in 1922, as cited in: Pullman, *The Atom in the History of Human Thought*, 262–263.

For example, Louis de Broglie posed the problem that these jumps could not be described in space, and Erwin Schrödinger argued that the idea of quantum jumps was nonsense. On the other hand, Bohr's hypotheses allowed him to calculate the radius of the hydrogen atom, its ionization potential, and to correctly reproduce the Balmer spectral series of this atom. Arnold Sommerfeld eventually replaced the circular orbits with the elliptical ones and introduced a second quantum number, then called *azimuthal*, to direct the main axis of the ellipse. To explain the effect of the magnetic field on the spectral lines, a third quantum number was introduced, the *magnetic* quantum number. These three quantum numbers could be related to the size, shape, and spatial orientation of the corresponding orbits.

In the case of the hydrogen atom, there was only one electron and this obviously occupied the lower-energy orbit. In the more complex atoms, the question arose regarding the maximum number of electrons that an orbit could contain. This problem was present in Bohr's mind, but it found a solution only in 1926 with Wolfgang Pauli's introduction of a fourth quantum number, that of *spin*, which in the case of the electron could have only two values. Pauli also introduced the principle of exclusion, which asserted that there are never two electrons with the same four quantum numbers. Therefore, the most that could be placed in an orbit are two electrons with a quantum number of opposite spin.

Meanwhile, the structure of the atom was enriched by another component, the neutron. In 1932, James Chadwick found a particle, in the atomic nucleus, with a mass almost equal to that of protons but without any electric charge. This particle allowed him to explain the discrepancy between the mass of the atom and the mass obtained by the sum of only protons and electrons. This particle also explained the existence of isotopes, which are atoms of the same element that exhibit almost the same chemical properties but have different mass because they differ by the number of neutrons. Isotopes were discovered by Frederick Soddy in the series of radioactive elements and then by Francis Aston in many other elements.

With the discovery of isotopes, one of Dalton's main assumptions that, for each element there is only one type of atom, was dismantled. The discovery of different types of water, each with a different molecular weight, had been rejected by Dalton on the basis of this flawed principle: "We can then conclude that the ultimate particles of all homogeneous substances are perfectly identical in weight, shape, etc."[6] Still in 1914, J. J. Thomson repeated almost the same words, "all particles of a given substance have exactly the same mass . . .all atoms of a given element are identical."[7]

---

[6] Dalton, *A New System of Chemical Philosophy*, as cited in *The Atom in the History of Human Thought*, 270.

[7] Thomson, as cited in *The Atom in the History of Human Thought*.

The Concept of Atom in Quantum Mechanics

As previously stated, the atomic vision at the beginning of the 20th century seemed to have reached a certain stability until everything changed again. In 1924, after radiation had been attributed to wave-particle duality, Louis De Broglie also extended this duality to matter. De Broglie himself said that this was where quantum mechanics was born. He wondered whether wave-particle duality reflected the deep and hidden nature of the quantum of action and whether one should expect such a duality wherever Planck's constant was involved. The role of integers in characterizing stable electron states in atoms was symptomatic, as integers are frequently involved in all branches of physics that deal with waves. It was, therefore, natural to suspect that the quantization rules would reflect some wave properties of electrons in atoms. One of the most interesting consequences of *de Broglie's relation*, which associated with each particle of momentum p a wavelength h/p with h Planck's constant, was that, when it was applied to electrons in an atom, it provided an elegant interpretation of the rules of quantization. Quantum numbers showed that the wave associated with a stationary state of an electron in an atom was itself a stationary wave, in the sense of conventional wave theory.

The impact of de Broglie's formula on the nascent quantum mechanics was rapid. Two years later, in 1926, Erwin Schrödinger[8,9] created that branch of physics known as wave mechanics. The basis of this new mechanics was the wave equation or Schrödinger's equation. In wave mechanics, the wave functions of electrons were called orbital and took the place of the orbits of the Bohr model. Schrödinger considered these wave functions to be as "real" as the acoustic waves. He thought they represented physical vibrations occurring in all directions of space. The main difficulty was that the wave function was described by an imaginary number, which made it difficult to associate it with a real vibration.

A completely different interpretation of the meaning of the wave function was proposed as early as 1926 by Max Born,[10,11,12] who introduced a probabilistic point of view of the subatomic universe. Leon Lederman, a modern high-energy physicist, considers this interpretation as "the most dramatic and profound change in our perception of the world since Newton."[13] Taking into account the

---

[8] Schrödinger, "Quantization as a Problem of Proper Values."
[9] Schrödinger "Über das Verhältnis der Heisenberg-Born-Jordanschen Quantenmechanik zu der meinen."
[10] Born, "Zur Quantenmechanik der Stossvorgänge."
[11] Ibid.
[12] Born, "Zur Wellnmechanik der Stossvorgänge."\
[13] Lederman and Teresi, *The God Particle: If the Universe Is the Answer, What Is the Question?*, 171.

complex nature of the wave function, the Born Principle states that the square of the wave function module at a given point and instant is a measure of the probability that the corresponding particle can be found at that point and at that instant. This probabilistic interpretation greatly changes our "normal" way of understanding the localization of a particle. It can no longer be said that, at the instant t, the particle is at point P with velocity v but only that there is a certain probability that this will happen.

In practice, if the velocity is fixed, the particle has indeterminacy in space and, if the position is fixed, there is no possibility of determining with "certainty" its velocity. This is a statement of the Heisenberg Uncertainty Principle, and it was a radical break with Schrödinger's interpretation that, in a classical way, we can understand the wave function as a description of a real material wave. To visualize the Born Principle, chemists think of the electron as dissolved in an electron cloud. The density of this electron cloud at a given point corresponds to the probability of the presence of the electron at that point, that is, to the square of its wave function. This is obviously just a graphical mental representation designed to visualize the probabilistic aspect of the wave function.

In Born's interpretation of the wave function, the probabilistic aspect is very different from that of classical probability as it was applied, for example, in the 19th century to the theory of gases. In the latter case, the system was considered entirely deterministic, and it was only our inability to master the enormous amount of trajectories, due to the large quantity of atoms present even in small portions of gas, that made a statistical approximation necessary. In quantum mechanics, the situation is completely different. In philosophical terms, for Born, the probabilistic aspect is ontological rather than merely epistemological. That is, it is a feature of the "nature" of the particle itself, not of our inability to know its trajectories.

The work of Werner Heisenberg had a considerable effect on this same issue. In 1925, Heisenberg proposed an alternative approach to Schrödinger's wave mechanics, known as matrix mechanics. We will not go into detail here regarding this type of mechanics, but we will emphasize only two of its features that are those characteristics relevant for our discussion. The first feature is that Heisenberg presented this mathematical formalism as a formal alternative to the "realistic" descriptions of atoms. The second feature is Heisenberg's elaboration of the uncertainty principle, which states that the product of uncertainty about position and momentum (related to speed) cannot be smaller than the Planck constant divided by 2π.

Although this principle can be seen as an algebraic consequence of Heisenberg's mathematical approach, it can be introduced in a more "empirical" way, as Heisenberg himself did, by means of a thought experiment. Suppose that we want to determine as accurately as possible the movement of an electron in an atom, that is, its position and its momentum. For this purpose, we decide to use a microscope. We need a microscope that uses a radiation of small wavelength, for example, the photons of gamma rays. We know that the position and the momentum of the electron are perturbed by the impact of the photons. The more precisely we attempt to determine the position of the electron, for example by increasing the frequency of radiation, the more the photons gain energy and the more the momentum of the electron varies, and vice versa. This problem cannot be overcome by changing the measuring instrument, and it imposes a limit on the system's response to our measurements. This limit is ontological; that is, it is intrinsic to the properties of the world of elementary particles. In quantum physics, the measurement of one variable can reduce the possibility of knowing another. The philosopher Gaston Bachelard notes that, before Heisenberg, errors on two independent variables were thought to be themselves independent, while now we understand that there may be a correlation between errors.

Comparing the classical treatment that we considered at the beginning of this chapter with what quantum mechanics establishes, we see that the conceptual "revolution" introduced by the latter is significant. This revolution does not merely refer to the quantization of some properties, highlighted by the term *quantum* used alongside *mechanics* in today's description of the atomic world. Certainly, the existence of "quantized" properties, that is, of properties that can only take some values and not all of the infinite intermediate values between them, is one of the "novelties" of quantum mechanics. But it is the existence of a quantum of action that also surprised its inventor, Max Planck, who also understood its disruptive action.

In his 1918 Nobel Lecture, Planck stated:

Either the quantum of action was a fictional quantity . . . or the derivation of the radiation law was based on a sound physical conception. In this case the quantum of action must play a fundamental role in physics. It was something entirely new, never before heard of, which seemed to be called upon to basically revise all our physical thinking, built, as this was, since the establishment of the infinitesimal calculus by Leibniz and Newton, upon the acceptance of the continuity of all causative connections. Experiment has decided for the second alternative.[14]

Heisenberg pointed out the close connection between the limited applicability of mechanical concepts and the separability of microscopic entities, both from the other microscopic entities and from the measuring instrument.

Alongside the "quantization" of certain properties, quantum mechanics also calls into question what, for millennia, had been considered the underlying

[14] Planck, *Nobel Lecture.*

hypothesis of all scientific endeavors, that is, determinism. Determinism was best described by Pierre Simon de Laplace's famous claim that an intelligent being who could know all of the forces of nature that are active at a given moment in time and the position of each entity at that moment in time would be able to predict the position of all entities at any given future moment in time. Such a being would be able to encompass in a single formula all of the movements of all things, from the largest physical body in the universe to the lightest atom. Nothing would escape its understanding, and the future would be immediately present before its eyes, just like the past.

Heisenberg, on the other hand, introduced a "quantum fog" since, if we cannot simultaneously determine the position and momentum of a particle at any given instant, we can never predict the future absolutely. However, the philosophical problem is even greater than this because, if we cannot simultaneously determine both the position and the momentum of a particle with infinite precision, we cannot affirm that the particle has both of these properties at the same time. These are the problems that quantum mechanics and its many interpretations have addressed and that lead to the problem of the "ontological" reality of the microscopic world. These fascinating topics, however, are beyond the scope of this book.

The dual wave-particle nature of matter has, however, also highlighted a further philosophical problem. As many authors have stressed, concepts like waves and particles that had been considered "absolute" and mutually exclusive have been relativized. However, wave-particle duality has also modified, in a more subtle and less evident way, the idea of matter. In classical mechanics, physicists and philosophers have traditionally adhered to the so-called substantive concept of matter. However, the wave-particle perspective of matter deprives matter of its intrinsic "substantive" nature.

Electromagnetism had already shaken this philosophical assumption since, although the charge has at least part of the mass function, its physical activity no longer resides in bodies but in the surrounding medium. Thus, the electromagnetic field is the seat of energy, and matter ceases to be the medium for physical events. Electromagnetism, in fact, was not only one of the first theories regarding fields, in the modern physical sense of the term, but has also fully expressed a fundamental principle of modern physics and now also of quantum physics. However, once microscopic entities are interpreted as being waves as well as particles with mass, the supremacy of substance in the microscopic world must be completely abandoned. The resulting philosophical conception of matter is that matter does not act as it does because it is what it is, but it is what it is because it acts as it does.

Differences and Similarities between Classical and Quantum Atoms

At the end of this overview, which has traced the new characteristics of the microscopic atomic world, we must pose a persistent question,[15] that is: can classical Greek atomism truly be considered the precursor of today's atomic theory? Surely the answer is positive. However, even a negative answer to this question would be correct to some extent. The atomistic theory of Leucippus and Democritus is both actually and historically configurable as the true precursor of today's atomic theory, but only if we take note that over the centuries the scientists responsible for the atomic vision were, ideally and concretely, looking back to those ancient Greek atomists. The atomic theory of Leucippus and Democritus is also, however, so different from contemporary atomic theory, and there are so many points of opposition between these two atomic visions, that we must question that ancient Greek atomism is the progenitor of current quantum atomic theory.

Let us now examine in detail the characteristics of the classical Greek atom that have been preserved in the contemporary quantum atom, as well as those characteristic that have been replaced. The Democritean atom may be schematized as follows:

1. The atom (from the Greek *a-tomos*"; that is, "nondivisible") was the elementary particle of matter. Macroscopic objects were more or less stable but never indivisible aggregates of indivisible parts, which were eternal, immutable, without parts or motion within them, and infinite in number. Indeed, the term *atom* indicated an essential property of these corpuscles, their indivisibility. However, other properties, such as the qualitative equality of all atoms and their intrinsic immutability, were equally characteristic.

2. The atom was indivisible because it did not contain within it any vacuum, which also made it impenetrable. Democritus said that, when a knife was used to cut an apple. for example, the knife had to find empty spaces in which it could penetrate to perform its function. If the apple did not contain empty spaces, it would be infinitely hard and, therefore, physically impenetrable and indivisible.

3. Atoms had different sizes, shapes, and movements. Atoms were fragments of the same being, the raw material that could not be further defined, and differed only in form, size, and motion. Atoms were all qualitatively identical, or, rather, they had in themselves no qualities other than those

---

[15] Pullman, *The Atom in the History of Human Thought*, 332–333.

mentioned above. These qualities were, at least in theory, quantifiable and mathematically identifiable. In fact, a very important aspect of the atomist program was the reduction of the qualitative changes of the macroscopic level to the quantitative changes of the atomic level. The atomists agreed with the Platonists and Pythagoreans that natural explanations had to be formulated in terms of geometric and numerical relationships. The "true" reality was only quantitatively accessible. Thus, true science was a mathematized science of nature.

4. According to ancient atomic theory, the qualitative differences between the various composite substances were attributed to differences in shape, size, position, and distribution of atoms. Classical atomism, by today's standards, provided an ad hoc explanation of macroscopic properties because it was not possible to establish its validity. Consider the dissolution of salt in water. The explanation hypothesized by the classical atomists was that this effect could be produced by the dispersion of salt atoms in the liquid. The classical atomists, however, could not explain why salt melted in water, while the same thing did not happen for sand. Of course, they could stipulate that the atoms of water entered the interstices between the atoms of salt but not those of sand. However, such an explanation, by today's standards, must be discarded because it is simply another way of saying that salt dissolves in water and sand does not. It is, therefore, a fundamentally circular explanation.

5. According to ancient atomic theory, all of the transformations that we observe in physical bodies were caused by the movements of atoms. In the process of aggregation, atoms did not lose their identity. Rather, they remained in contact with each other and were juxtaposed. Each atom was immutable, and the only thing that atoms did was to move and collide with each other and, if their forms made this possible, they combined with each other. Aristotle criticized atomism precisely because these atoms could not form really "new" entities when they formed aggregates. Galen rejected atomism for the same reason. At the birth of modern chemistry, Lavoisier applied this same principle to the elements, and his Law of Conservation of Mass reaffirmed this concept: "Nothing is created, neither in the operations of art nor in those of nature, and it can be established in principle that in every operation there is an equal amount of raw material and after the operation itself; that the quality and quantity of the elements is the same, and that there are only changes and modifications."[16]

In contrast, the quantum theory of atoms may be schematized as follows:

[16] Lavoisier, *Traité Élémentaire de Chimie*.

1. The quantum atom is no longer the elementary particle of matter. It consists of a nucleus, which contains protons and neutrons, and of electrons. Electrons are elementary particles and, together with muons, tau, and three types of neutrinos, form the family of leptons. Protons and neutrons belong to the family of hadrons and are not elementary particles. Like other particles, they consist of the six types of quarks. The six leptons and six quarks form the standard model of the universe. These particles, however, are only part of the current picture of elementary particles (the so-called fermions) because it is still necessary to add the particles that correspond to the forces of interactions (the bosons), the prototype of which is the photon for the electromagnetic field. The current corpuscular physical vision presents twelve elementary particles that form matter and twelve elementary particles that transmit force. Once we include the antiparticles and the differentiation of the quarks into "colors," we arrive at sixty types of elementary particles.

2. The quantum atom is neither indivisible nor impenetrable and consists largely of mostly empty space. Apparently, this is the greatest novelty of the modern atom, which makes the name itself contradictory precisely because it is no longer an indivisible particle. In reality, this difference for chemistry is unimportant (with the exception, of course, of radioactive elements). According to Heisenberg, if the term *atom* indicates indivisibility, then it was more suitable for protons, electrons, neutrons, mesons, and the like, as these were the "elementary particles" of his time. For this same reason, some contemporary physicists may consider the term *atom* suitable for quarks and other contemporary elementary particles. For contemporary chemists, however, the elementary particle is and remains the atom. This position was well expressed by Kekulé in the 19th century: "Even if scientific progress were to lead someday to a theory of the constitution of chemical atoms—as important as such as an advance would be to a general philosophy of matter—it would make little difference in the field of chemistry. The chemical atoms will forever remain a chemical unit."[17] Moreover, removal of the property of impenetrability allowed overcoming what for centuries had been the main conceptual obstacle of the atomic vision: the formation of aggregates. At the moment of their association, they "interpenetrate" in the sense that the valence electrons (or at least some of them) become the common patrimony of the molecule.

3. With regard to contemporary atoms, we can still speak of different sizes if we mean by this the average position of the electrons that are farthest from the nucleus. All electrons have an average distance from the nucleus, the

---

[17] Kekulé, as cited in *The Atom in the History of Human Thought*, 232.

old Bohr orbit, which is the function of the principal quantum number. Generally, the higher this quantum number is, the more these electrons are "averagely" further away from the nucleus. Therefore, an atom with many electrons, which then fills the shells of valence with a high principal quantum number, has a greater radius in this sense. Of course, these atomic quantities are not infinite in number. Referring to classical magnitude, one could say that the atom of each element (and of each ion) has a fixed size. We see that, although in a certain sense we can still speak of "atomic ray," we are dealing with concepts different from the size of the corpuscles postulated by the ancient Greek atomists. With regard to form, this concept has lost the importance that it had for the ancient Greeks. The atomic orbitals still have their own "shape," whose dimensions are certainly not infinite. However, the concept of atomic form has lost all meaning, though, as we shall see, "form" will still be important within the microscopic world of molecules and macromolecules.

4. Today's atoms are qualitatively different from each other. For each element (excluding isotopes), there is only one type of atom, and this is a nontrivial and nonexplainable phenomenon in classical mechanics. Further, the atoms of one element are qualitatively different from those of another. Dalton differentiated them by weight, and today we differentiate them by their constituents and their structure. In the next several chapters, we will examine in greater detail the importance of the "structuring" of matter and its relationship to "qualitative" differentiation.

5. What we observe in composite bodies depends as much on their different nature as on their movements. But, above all, it depends on their aggregations. This point is essential to the perspective taken in this book. In fact, the need to focus on a molecular theory of matter, as distinct from the atomic theory, comes from these last two points. The qualitative differences of the composite substances and their different transformations depend, of course, on the atomic aspects. However, the aggregates of atoms form a unique whole (the molecules) that is so new that it needs a proper name and becomes a new subject of action to which the macroscopic properties refer.

These five points illustrate what was already discussed, that is, although the ancient atom is actually the historical precursor of the modern atom, it is still very different from it. However, what allows us to reconnect to the ancient version of this concept to the contemporary version is the fundamental idea that the microscopic world is in close relation to the macroscopic one and to his transformation. This idea, introduced in antiquity, received confirmation in later centuries, when what was fundamentally a philosophical idea very gradually became a scientific theory.

We have seen how the atom has changed in contemporary physics. However, we must still consider the transformation of another essential element of classical atomism, that is, the vacuum. Quantum mechanics has substantially changed this concept by means of Heisenberg's Uncertainty Principle and Einstein's equivalency of mass and energy. Without going into great detail, we define the vacuum in quantum mechanics as the state of minimum energy of matter (of all fields associated with it) and as an "ocean of virtual particles" that form and dissipate. It becomes an energy bank and, due to the possibility of interconversion between mass and energy, it also becomes a particle bank. Emptiness becomes a virtual state of the universe, so that the vacuum and elementary particles become two different manifestations of the same "reality." It also participates in the properties of matter, and we can talk about polarization of the vacuum, which yields a fusion of two terms that would have been opposed in classical atomism.

The Chemical Bond and the Molecule in Quantum Mechanics

Quantum mechanics, by changing the concept of atom, has also given a less model sense to the concept of chemical bond, namely, the "glue" that holds the atoms together. The first chemical system studied by the new quantum mechanics was, of course, the hydrogen molecule. In 1927, Fritz London and Walter Heitler determined the energy of H_2. In this system, two electrons hold together two nuclei, each consisting of a proton. When the two nuclei are distant and each has an electron with it, the energy of the system equals that of two isolated hydrogen atoms, and only the force between the electron and the proton must be considered for each atom. When the two nuclei are close together, four interaction forces are involved. With the help of the system used by John William Strutt, Lord Rayleigh to estimate the minimum energy of bell vibrations and electron indistinguishability, they discovered a simple function that, when inserted into the Schrödinger equation in order to obtain the wave function, connected to the molecular electronic distribution.

In this way, a binding energy resulted that was very similar to the one discovered experimentally by using spectroscopy. This was the first step in the development of quantum chemistry. The object of study of this branch of chemistry is the application of quantum mechanics to molecules. The existence of compounds has finally become "comprehensible," and the fact that electrons do not belong to the individual atoms but to the entire molecule makes the conception of a molecule as a summative aggregate of atoms no longer viable. That is, the molecule as a whole can no longer be conceived merely as an aggregate summation of its atomic parts.

Obviously, the situation is far from simple. Even if we want to overlook the enormous "technical" difficulties of the study of molecules, the physical perspective of "molecular" electrons collides with the chemical perspective of bonds as a privileged relationship between pairs of atoms. Without wanting to delve into the largely unresolved problem of the application of quantum mechanics to chemical systems, we will now focus on the quantum concept of orbital that replaces the classical concept of orbit in atoms and molecules. By the late 1920s, two types of atomic orbitals, called s and p, had been discovered . In polar coordinates, the solution of the Schrödinger equation for one electron and an atomic nucleus was found to be a symmetric function called orbital s. The subsequent electronic state was found to be three times degenerate, with the symmetry of one of the three Cartesian axes, and was called the atomic orbital p.

The chemical bond between two atoms was explained qualitatively as an overlap of these atomic orbitals. Hund's rule of maximum multiplicity established that electrons occupied as many orbital atoms as possible to avoid mutual repulsion. In the case of water, Linus Pauling established that, since oxygen has four electrons in the last occupied electron shell (which is called valence and is identified by the principal quantum number equal to 2), by Hund's rule one of the p orbitals was doubly occupied, and the other two contributed two bonds to the hydrogen's orbital. The problem was that, in this case, the water would have to form an angle of 90° between the two O-H bonds. This problem was resolved by postulating repulsion between the two bonds.

Unfortunately, this schematization of chemical bonds in molecules did not seem to work for the carbon atom. This atom, with two fewer electrons than oxygen, only has two p electrons in the valence shell, and, therefore, its tetra valence remained a problem. In 1928, Pauling created the concept of "hybrid orbital" to overcome this issue. If we assume that one of the two 2s electrons is promoted in an orbital p, we now have four unpaired electrons in the valence shell. Pauling showed that the energy required to promote electrons from a 2s to a 2p was largely obtained by allowing carbon to form four bonds instead of two. For the four bonds to be equivalent, the four orbitals (a 2s orbital and 3 2p ones) had to combine to form four sp^3 hybrid orbitals. Pauling theorized the existence of other types of hybrids to explain other obtainable geometries. Since, after a few years, John Slater mathematized this approach, it was named, in the 1930s, the Heitler–London–Slater–Pauling method. Later, the name was changed to Valence Bond (VB).

Pauling introduced another important concept in his book, *The Nature of the Chemical Bond*, which was the electronegativity of an atom, defined as the tendency of an atom to attract to itself electrons shared in bonds. An alternative method of describing the chemical bond as a global combination of atomic orbitals was the molecular orbital approach (MO), introduced in 1927–1928 by Friedrich Hund and Robert Mulliken. In the 1940s and 1950s, this method, along with the contributions of Charles Coulson, Christopher Longuet-Higgins, and Erich Hückel, was adopted in theoretical chemistry in the interest of providing greater simplicity of calculation.

This method was based on the fundamental idea that, as one could describe an electron in the atom with its atomic orbitals, it was also possible to describe an electron in the molecule with the help of another function called the molecular orbital. The only difference between these two types of orbitals was that the first was monocentric, that is, centered on one atom, whereas the second was multicentric because each electron was subject to the simultaneous action of all nuclei. The goal of this method was to determine the molecular orbital for each electron in the molecule and, in particular, its energy.

As in the case of an atom, the square of the molecular wave function module represented the probability of finding the corresponding electron at a given point of space and time and, more immediately and globally, the shape of the associated electronic cloud. Still, as in the case of the atom, each orbital could "host"[18] at most two electrons of opposite spin. As in the atom, the orbitals were filled in the energetic order until the electrons ran out and the remaining orbitals, called virtual, could be occupied in electron excitation processes. Also in its standard form, molecular orbitals were assumed to be a linear combination of atomic orbitals (LCAO approximation) and, because molecular orbitals had symmetries, they were again in analogy with atomic orbitals.

One of the main difficulties of the VB method was the problem of resonance formulas. The analogy with the classical case, in which the concept of resonance was related to the separation of two systems that had the same frequency and that interacted, led to the idea that there really were electrons that fluctuated between two possible configurations of equilibrium. Pauling himself[19] encouraged the view that the A-B bond was obtained by the actual overlapping of ionic and covalent forms. This was certainly the way in which the Marxist School interpreted Pauling's thought and began to attack him for the deformation he had created in the concept of molecular structure (referring to the Russian chemist Alexander Butlerov) and for his creation of "fictitious molecules." James Steuart Dewar expressed well the view that was predominant in the field of theoretical chemistry by asking why it was necessary to complicate matters by representing benzene as the mixture of so many formulas of structure, not to mention the more

[18] Villani, Ghibaudi, and Cerruti, "The Orbital: A Pivotal Concept in the Relationship Between Chemistry and Physics? A Comment to the Work by Fortin and Coauthors."
[19] Pauling, *The Nature of the Chemical Bond*.

complex compounds in which a huge number would be used. The MO method was much simpler, and, therefore, it was this method that must be implemented.

Characteristics of Contemporary Chemical Bonds

Let us now summarize the main characteristics of interatomic bonds (called chemical bonds), as conceived by current quantum chemistry. All the bonds between atoms are of an electrical nature because, among the fundamental physical forces, the electromagnetic is almost entirely responsible for chemical bonds.[20] Two large classes of interatomic bonds are distinguished: ionic and covalent, but these two classes are ideally connected by all intermediate cases. Ionic bonds are more relevant for inorganic chemistry, while covalent bonds are more relevant for organic chemistry. Ionic bonds are formed by the practically complete transfer of an electron from one atom to another and by subsequent reciprocal attraction between these two units with opposite charges. Covalent bonds are formed by a symmetrical or nonsymmetrical sharing of a pair of electrons. The following characteristics apply mainly to covalent bonds:

1. Various types of bonds (σ, π, etc.) are differentiated by symmetry. As is true of atomic orbitals, molecular orbitals that produce chemical bonds have characteristics that can be linked to spatial symmetry.
2. The formation of chemical bonds can be seen in terms of interaction between particular combinations of monocentric atomic orbitals (hybrids) rather than simple atomic orbitals. Sometimes, hybridization is almost considered a sort of "preparation" of the atom for the formation of the bonds between the atoms. Obviously, this "temporal" vision is not correct, and the hybrid orbital, if one wishes to use this model, is formed gradually as the atoms that must bind together come closer.
3. There are substantially localized bonds in which a pair of atoms is involved and delocalized bonds involving many atoms. The molecule of benzene, with the π orbitals delocalized on the six carbon atoms can be an example of the delocalized bond, while, again in the molecule of benzene, the C-H bonds are localized.
4. A bond usually involves two electrons. However, in some cases it involves only one electron (H_2^+, for example).

---

[20] The authors are aware that, in the 1970s, Sheldon Glashow, Steven Weinberg, and Abdul Salam discovered that the electromagnetic and weak forces, which appear different at everyday low energies, were two aspects of the same electroweak force at very high unification energies (the Shaldow–Weinberg–Salam theory). However, this unification is not relevant for the energy levels involved in chemical bonds.

5. In metals, the bonds are completely delocalized, transforming the discrete energy levels of the molecules into a continuous range of energy (bands).
6. There are several different types of intermediate energy interactions between common chemical bonds and physical forces. An example is the hydrogen bond, which is formed between a hydrogen atom, bound to an electronegative atom, and another electronegative atom of the same molecule or another molecule. Such bonds are responsible for the "liquidity" of water at room temperature and for many biochemical properties.

In conclusion, today's quantum atom is a scientific concept that is difficult to visualize and, therefore, difficult to render "comprehensible" for both the specialist and the layperson. Here, we have emphasized the diversity between the ancient and the contemporary atom, focusing mainly on the general implications underlying the two atomic visions. The atomic vision that persists from Democritus (or Leucippus) to Newton (with the partial exception of Boyle) is a classic, paradigmatic, reductionist vision of reality. What reality is and what it is transformed into are both explainable in terms of its fundamental constituents, that is, atoms and their interactions. This point should be clear enough so that no further clarification is required at the moment.

The contemporary atom has evolved from Dalton to the present day and, therefore, does not merely encompass the quantum atom, which implies an entirely different philosophical vision. In Chapter 6, we saw that Dalton considered atoms of different elements, that is, chemical atoms, as being qualitatively different. The quantum atom also has this essential property, but this property is linked to its structure, to its being a system. The fact that some of its properties are comprehensible in terms of its constituents does not in any way eliminate the need, in general, to consider the atom as a global entity. This is true for atoms and is even more true for molecules and macromolecules, and this goes back to the general concept that a "structured reality" constitutes a system in the "technical" sense that we are elaborating in this book.

A fundamental characteristic that differentiates atoms is their different ability to interact with other atoms. Neither Newton's law of attraction nor Coulomb's law of electrostatics explains the interaction between atoms, and the problem of explaining this "specificity" has persisted throughout modern scientific history. Current quantum mechanics gives a "complex explanation" of the chemical bond. We believe that the molecular structure, intrinsically linked to the concept of chemical bond, constitutes an "emerging property" in the sense meant by the sciences of complexity and systemics. Therefore, from this systemic and complex perspective, a transformation of a chemical body (that is, a chemical reaction) consists in its deconstruction and subsequent restructuring.

11

The Emergence of the Concept of Macromolecule

The concept of the biological macromolecule emerged in the first half of the 20th century. This concept considers the existence of giant molecules with a molecular weight of tens of thousands or even millions of daltons (a dalton, Da, is approximately the atomic weight of hydrogen). In the case of the macromolecule of DNA, this concept required scientists to overcome two distinct epistemological obstacles. The first of these obstacles is the possibility that such large molecules can be held together by normal chemical bonds. For a long time, the hypothesis that these giant molecules existed was opposed to the idea that such substances were "aggregates" of small molecules, as in the case of the colloidal nature of substances of living beings, which we will examine later.

The second epistemological obstacle is illustrated by the aperiodic polymer. Even when the existence of such an enormous molecule was believed possible, it could only be formed by a single unit that was repeated countless times, which is the case with symmetric polymers. Already in the 19th century, a macromolecular explanation for cellulose had been advanced, which considered it to be a sugar chain of indefinite length. As we will see, the concept of the aperiodic polymer, which is formed by different repetitive units, is one of the basic ideas behind the concept of the biological macromolecule, whether in proteins or in nucleic acids. Before analyzing these two general epistemological obstacles, let's consider a more specific case: a macromolecule that displays the characteristics of life.

The Protoplasmic Theory

The protoplasmic theory stipulated that the tissues of living beings were formed by an aggregate of small gelatinous masses, without membrane and without nucleus and that were not, configurable as cells. The protoplasm was considered a living form simpler than the cell but capable of realizing the typical functions of living beings. The term *protoplasm* was, therefore, considered to be synonymous with *living matter*. This theory did not connect life to an organization (as in cell theory) but to this special matter called protoplasm. As Haeckel tells us: "Life, properly speaking, is linked not to a body of a certain form, that

From the Atom to Living Systems. Marina Paola Banchetti-Robino and Giovanni Villani, Oxford University Press.
© Oxford University Press 2023. DOI: 10.1093/oso/9780197598900.003.0012

is morphologically differentiated and equipped with organs, but to an amorphous substance of a physical nature and a determined chemical composition."[1] Chemistry could analyze and determine the proportions of the constituent elements of this substance because the atoms of this matter were ordinary ones. Life was not inherent in the organization of cells, tissues, and the like, but in the organization of atoms in these "albuminous" molecules ("albuminous" being the term then used to refer to protein molecules). Haeckel, therefore, belongs fully to the group of scientists and philosophers who have attempted to associate life with a "living" matter.[2]

We contend that this vision of the living world is grounded in a philosophical error that is typically reductionistic. This is the assumption that, by putting together nonliving things, one cannot obtain things that are alive and that therefore the constituents of living things must themselves be alive. This approach is inconsistent because it considers that, even in the case of "living molecules," these are formed by "normal" atomic constituents that do not have the property of being alive. Let us recall that, for the ancient Greek atomists, there was nothing to distinguish the atoms of animate matter from those of inert matter. It was Descartes who proposed a dichotomy of substances (*res extensa* and *res cogitans*) which, as we have seen, did not concern the atomic world and did not differentiate living from nonliving matter.

As was discussed extensively in Chapter 2, in order to oppose this dichotomy, the 18th-century atomists Moreau de Maupertuis and Diderot assigned to all atoms some principles of intelligence, aversion, memory, and the like. Indeed, they conceived of no other possible way for material physical forces to account for the manifestations of life. Diderot stated: "To suppose that by placing next to a dead particle one, two, or three other dead particles, one can form the system of a living body amounts, it seems to me, to a flagrant absurdity."[3] As Chapter 2 emphasized, there was, on the one hand, the need to account for life, and, on the other hand, there was the need to explain how inanimate food could become part of living bodies. These questions led Maupertuis to postulate "living points" as the fundamental constituents of all living things.

The Colloidal State

Throughout the 19th century, some evidence had accumulated showing that complex organic molecules, such as proteins and nucleic acids, were composed

---

[1] Haeckel, as cited in André Pichot, *Expliquer la vie. De l'âme à la molécule*, 906.
[2] Ibid.
[3] Diderot, as cited in *The Atom in the History of Human Thought*, 148–149.

of smaller molecules that were not held together by normal chemical bonds. In 1861, Graham proposed the concept of "colloid" for proteins. He distinguished natural substances into two large groups: colloids (and, as an example, he reported gelatin) and crystalloids. Colloids were a particular form of aggregation that constituted the colloidal condition of matter. Crystalloids formed solids with well-defined faces and angles, while colloids formed amorphous solids that grouped small units of crystalloids. The father of chemical structuralism, Kekulé himself, believed that colloids were sponge-like masses, that is, lattices formed by small molecules.

The colloidal state was a viscous state that was intermediate between the solid and liquid state. It was also a dynamic state or energetic state, whose energy was dissipated in vital processes. Underlying this theory was a general problem regarding the properties necessary for life-supporting substances. On the one hand, there were solid-state materials. These had a fixed shape, but it was believed that they did not have time to change shape to be "fit" for life. On the other hand, there were materials in the liquid state. These did not have their own shape, and even their temporality was not considered "suitable" for life because all changes in form were too rapid, almost instantaneous. To delineate the activities of living matter, a substance with intermediate characteristics was needed, and this substance was found in colloidal materials. Colloidal matter had a certain shape that could be modified at any time and slowly enough to be useful in describing the processes in living things. The colloidal state, therefore, reconciled the form and temporality of matter with the static and dynamic needs typical of living things.

Another essential point to consider was the relationship between life and the environment, and, again, the colloidal state seemed to have the right characteristics. The shape of the colloid was susceptible to transformation under the influence of different external actions (e.g., temperature). Colloidal matter, therefore, had a certain "sensitivity" to these external properties. Thus, everything seemed to connect colloidal matter to life. A further example of the relationship between the colloidal state and life were the slow reactions that served the latter and that were explained, in this theory, by the slow diffusion of reactive substances in colloidal matter. It was believed that, because of its slow diffusion, colloidal matter could "control" and "correlate them" to the other reactions, since their rate of diffusion could depend on other simultaneous reactions capable of modifying the state of the colloidal substances.

Simply put, the solid state was thought to be too organized and too rigid to explain life. The liquid state, on the other hand, was too disorganized and too fast to be modified organically to serve life. Finally, the colloidal state was suitable for life because matter in that state was neither too organized nor too disorganized and was the location where vital processes could interact through internal and external physical and chemical changes. The problem that the theory of the colloidal state sought to resolve was, therefore, to find a material in which organization could occur, a problem that still needs to be addressed. To explain life, it was necessary that the matter of living beings be "soft" but also susceptible to organization, in order to give life to organized beings. The colloidal state had the advantage of reconciling the fluid state necessary to the chemical environment with the solid state that was required for organization. The solution was seductive and highlighted problems that are valid even today. In order for macromolecules to be useful in understanding life, it was necessary to theorize that they were not an indistinct and gelatinous substratum, but that each of them has its own well-defined shape that could also change under the action of different agents.

As we shall see, the shape of macromolecules and their "sensitivity" to physical and chemical agents played a role in the development of cellular chemistry, a role similar to that played by colloidal substances. From the beginning, the study of living matter was confronted with the problem of how chemistry could explain life, since chemistry was extraneous to the concept of organization, which is essential to the explanation of life. This problem was also related to the idea that chemistry always required a statistical approach in order to study reactions. The reactions that occurred in living beings were different from those in the gas phase, since they were individual nonaverage statistical events. We will later return to this point, which was also essential for Schrödinger in his book *What Is life?*

Toward the Macromolecule

The early 20th century marked the widespread extension of Kekulé's normal valencies from molecules with a few atoms to giant molecules. Hofmeister and Fischer's theory of polypeptides postulated a substrate common to the concepts of molecule and macromolecule by proposing that proteins were polymers of amino acids (small molecules) that were bound together by normal chemical bonds, that is, the amide bonds. In 1902, Franz Hofmeister stated that "[o]n the basis of these given facts one may therefore consider the proteins as for the most part *arising by condensation of amino acids, whereby the linkage through the group -CO-NH-CH=* has to be regarded as the one regularly recurring."[4] For Emil Fischer, however, the maximum polypeptide size that could be obtained with peptide bonds was about 40 amino acids, corresponding to a molecular weight of about 4000–5000 Da. The natural proteins that had a higher molecular weight were, therefore, a mixture of these polypeptides.

---

[4] Hofmeister's lecture at the *Geselhchaft fur deutscher Naturforscher und Arzte*, as cited in Graeme K., *Vital forces: The Discovery of the Molecular Basis of Life*, 118.

At the same time, Thomas Osborne and Isaac Harris considered that nucleic acids were also huge molecules in which nitrogen bases, sugars, and phosphoric acids were somehow bound together: "To the polyphosphate backbone are attached three molecules of pentose ($C_5H_9O_5$), one molecule of guanine ($C_5H_4N_5O$), one molecule of adenine ($C_5H_4N_5$), two molecules of uracil ($C_4H_3N_2O_2$) and 'an unidentified basic product' (X)."[5]

In 1911, Albrecht Kossel elaborated the concept that the complex molecules of living beings were composed of "*bausteine* [building blocks]," simple organic molecules that could be combined in different ways:

So long as one considers the mass of living substance as a whole, an analysis of its activity can scarcely be undertaken. Such an analysis is only possible through the isolation of certain units capable of chemical investigation and to whose activity the individual functions of living substance may be referred. I wish to speak of these units, which I shall refer to as the "*Bausteine*" or building-stones. . . . Through the union of these Bausteine larger aggregates are formed which we call either proteins, fats, nucleic acids, phosphates, or polysaccharides, as the case may be.[6]

From the historical point of view, the *baustein* hypothesis recognized that organic molecules could be combined in different ways to produce the molecules needed for life, although Kossel, like Fischer, believed that macromolecules were probably aggregated from smaller polymers. Kossel also understood that, during digestion, proteins were split into amino acids and that these were then reused to form new proteins. The *baustein* hypothesis also represented the decisive step toward the aperiodic polymer, and this happened not just for one class of substances (proteins or nucleic acids) but for many classes (proteins, fats, nucleic acids, and polysaccharides). In 1917, Søren Sørensenn showed that by dissolving egg albumin in water, a solution was formed rather than a suspension, as had been expected if it were an aggregate of polypeptides of different sizes, and he calculated a molecular weight of 34,000 Da for this protein.

Among the experimental techniques used by biochemists at the beginning of the 20th century were spectroscopy, which used ultracentrifuge and the newly discovered X-rays. The second of these techniques revolutionized the analysis of molecular structures. The chemical analysis methods used until that point had given the composition of a substance and, only in a very indirect way, had identified the localizations of atomic groups within the molecule. The spectroscopy that was already in use often gave ambiguous results that needed

---

[5] Osborne and Harris, as cited in *Vital Forces: The Discovery of the Molecular Basis of Life*, 112.
[6] Kossel, "The Chemical Composition of the Cell."

interpretation. But X-ray spectroscopy offered the possibility of locating the position of atoms in crystals. As William Bragg would state in 1915, we find ourselves able to measure the actual distances from atom to atom as if we were tracing the blueprints of a building.

There were, of course, limitations to the applicability of this spectroscopic technique. The first limitation was that such a technique required the test substance to form a crystal. This difficulty was overcome in biochemistry by the discovery that even the fibers of biological substances were sufficiently periodic and, because they could practically be considered a monodimensional crystal, they were amenable to X-ray spectroscopy. The second limitation was the considerable complexity of the spectrum as the molecular complexity grew. This difficulty was partially overcome with the practice of analyzing experimental results.

The ultracentrifuge was invented by Theodor Svedberg (1884–1971), who also discovered the relationship between the rate of sedimentation and the molecular weight of a substance. After building this machine, which reached 42,000 rpm, Svedberg demonstrated conclusively that many biological substances were not, at the molecular level, aggregates of small molecules but rather were molecules with an atomic weight of tens of thousands of daltons. Hemoglobin, for example, had a molecular weight of 66,800 Da and corresponded to four subunits of 16,700 Da. A further blow to the colloidal state theory came from Hermann Staudinger (1881–1965) who, working on synthetic polymers, identified a relationship between their viscosity and the molecular weight of the polymer. This relationship is known as Staudinger's law.

With regard to macromolecules, the fundamental problem of reconstructing the order of monomers that made them specific remained unsolved at the end of the 1930s. In particular, it was necessary to determine the order of amino acids in proteins and to decipher the specific sequences in nucleic acids. Besides this "technical" problem, there was also a conceptual problem that needed resolving. The synthesis of a homopolymer (i.e., a polymer formed by a single monomer such as cellulose, which has glucose for a monomer) was conceptually easy. Such macromolecules required an enzyme capable of binding monomers. However, the synthesis of an aperiodic polymer with many types of monomers required not only many enzymes but also a "time method" to arrange the different monomers.

From the conceptual point of view, the idea was developed that the peptide chain of the protein had a natural form that was generated by the phenomenon of folding, and that the protein could lose its form through the so-called phenomenon of denaturation. However, weak intramolecular bonds, such as those of hydrogen, held this specific "molecular form" in place. In 1936, Pauling stated that the native form of a protein consists of one polypeptide chain which continues without interruption throughout the molecule . . . this chain is folded into a uniquely defined configuration in which it is held by hydrogen bonds. . . . The characteristic specific properties of native proteins we attribute to their uniquely defined configurations. The denatured protein molecule we consider to be characterized by the absence of a uniquely defined configuration. . . . It is evident that with loss of the uniquely defined configuration there would be loss of the specific properties of the native protein.[7]

It should be stressed that this is a philosophical point as well as a technical one. The macromolecule does not differ from the molecule by the simple fact of being much larger. It is also more complex, in the technical sense of the term. The molecule is determined by its structure, that is, by the internal organization of the relations between its constituents, which are linked by chemical bonds of "Kekulé"-type value. The properties are connected to this structure. The macromolecule goes beyond the molecular structure, in the sense that its properties are not determined by the molecular structure alone, that is by the specific chain or peptide chains, but also by its shape. The molecular structure, in fact, does not vary in the process of folding and/or denaturation of a protein, yet these two macromolecules are "very different.

We will return shortly to the difference between structure and shape in molecules and macromolecules, and we will also demonstrate how the problem of determining the sequence of monomers in both proteins and nucleic acids was resolved. However, for the moment, we will focus with some scientific detail on the process of protein folding. Our purpose here will be more philosophical than scientific, although the scientific details will serve to clarify the philosophical considerations, which are important in general and not only for the folding process.

In vitro and In vivo Protein Folding and Denaturation

A protein within a biological environment reaches its most stable three-dimensional conformation, that is, its lowest-energy state,[8] which is called the native state of a protein. It is the protein in this native form that is biologically functional. The process that leads the protein to take its native form is called

---

[7] Mirsky and Pauling, "On the Structure of Native, Denatured, and Coagulated Proteins."
[8] In reality, the minimum reached by a system is that of free energy and this depends on the energy and entropy of the state. In this context, however, it is not necessary to consider these "technical" issues.

the folding process. The process that leads the protein to the loss of this form is called, instead, the denaturation process.

The information necessary for a denatured polypeptide to reach a stable conformation is contained within its own amino acid sequence. Obviously, this stable configuration depends more or less on the environment in which the macromolecule is found. If this stable configuration is calculated for the isolated molecule (i.e., *in silico*), a certain result is obtained that can be different in vitro (the macromolecule in a tube) and in vivo (the macromolecule in a cell). In the calculation *in silico*, we have considered only one macromolecule. In the calculations in vitro, we also take into consideration the interactions between the specimens of the same macromolecule and their interactions with the solvent that is present. Finally, for calculations in vivo, we must consider the complex environment in which the protein exists. In all three cases, it is clear that the protein is able to spontaneously assume a stable three-dimensional structure. But this, however, does not mean that the biologically active conformation of a protein can be inferred simply from knowledge of its molecular structure (i.e., its primary structure).

In fact, in vivo, the process involved in the assumption of a correct steric conformation and the final achievement of its native conformation involves a complex intertwining of interactions among different enzymes and chaperones (a group of proteins that assist the protein in its folding process). It is generally believed that, in vivo, the folding of proteins proceeds through multiple steps. A large number of interactions must take place, both within the same folding chain and with specific folding enzymes, as well as with adjuvant proteins (chaperones). The folding chain must be formed and broken during this complex process. All this can lead to a stable configuration that is very similar to that which would be obtained *in silico* and in vitro or to a more or less different stable conformation.

This problem was clear to Christian B. Anfinsen, who won the Nobel Prize for Chemistry in 1972 for his work on the connection between the sequence of amino acids and the biologically active conformation of a protein (in his case of Ribonuclease). In his Nobel Lecture, he stated that

> [t]his hypothesis [the thermodynamic hypothesis] states that the three-dimensional structure of a native protein in its normal physiological milieu (solvent, pH, ionic strength, presence of other components such as metal ions or prosthetic groups, temperature, etc.) is the one in which the Gibbs free energy of the *whole system* is lowest; that is, that the native conformation is determined by the totality of interatomic interactions and hence by the amino acid sequence, in a *given environment*. In terms of natural selection through the "design" of macromolecules during evolution, this idea emphasized the fact

that a protein molecule only makes stable, structural sense when it exists under conditions similar to those for which it was selected—the so-called physiological state.[9]

This principle, later called Anfinsen's Principle, "asserts that all information required to specify the structure of a protein is encoded in its amino acid sequence."[10] The insistence on the environment, on "its normal physiological milieu," and on "conditions similar to those for which it was selected" has disappeared.

In practice, although the in vitro and in vivo protein may have, and often have, a stable configuration similar to the native configuration, this configuration depends both on the primary structure of the protein and on the environment in which it is found. From the thermodynamic point of view, this is obvious: a closed system is in its minimum (free) energy within an environment, and if the environment changes (in vitro and in vivo), the state of minimum energy changes and with it the form of the macromolecule. From Anfinsen's early experiments on in vitro refolding of denatured ribonuclease, it became clear that if the essential information contained within the same polypeptide chain is added to the "correct" environmental conditions for refolding (an in vitro environment similar to the in vivo environment), a denatured protein will reach its native conformation in the absence of any other substance. Of importance in achieving a proper folding are some key amino acids, the general nature of the amino acid sequence, and also some essential characteristics of the environment. This is the case because other substances that help the folding process can also be important, although they are not essential.

Many enzymes and other nonstructural proteins have more than one native state, and they operate in, and can be regulated by, the dynamics of the transition between these stable states. The term *native state*, however, is almost always used in the singular to contrast it with the denatured state and to highlight the difference between the protein in the active and nonactive forms. The case of proteins with an allosteric effect is the most evident case of proteins with different stable states. The allosteric site of a protein indicates the presence, even in a very distant position from the protein's active site, of another site that can regulate the activation of the protein, following the interaction with a molecule. In practice, the addition of a molecule that binds to the allosteric site changes the shape of the protein, thereby controlling its biological activity, although sometimes the process is more complex than the simple modification of the shape.

---

[9] Anfinsen, "Principles that Govern the Folding of Protein Chains."
[10] Sander, Chaney, and Clark, "Expanding Anfinsen's Principle: Contributions of Synonymous Codon Selection to Rational Protein Design."

The understanding of how proteins "fold" in each environment, thereby giving rise to a specific form, helps to clarify the intrinsic and extrinsic aspects of this process, the intrinsic aspects being those related to the genetic information that determines the monodimensional structure of the protein and the extrinsic aspects being those related to the environment. Moreover, the understanding of the mechanisms that determine the shape of the protein, and in particular its native shape in the cell, has also proved to be a practical tool of incalculable value. We recall that errors in protein folding have recently been associated with several diseases, such as *Amyloidosis* (a group of diseases in which abnormal proteins, known as amyloid fibrils, build up in tissue), since any error that leads to the loss of a properly functioning protein can lead to a dysfunction and a disease. Another example of a disease associated with incorrect protein folding is cystic fibrosis.

The types of errors that can occur in folding include the destabilization of the native state, that of an intermediate state of the protein in its folding path, the prolongation of the bond with a molecular chaperone or with an enzyme involved in folding, the formation of collateral preferential folding pathways that lead to nonfunctional protein situations and, finally, the folding in wrong intracellular compartments. Understanding the mechanisms of protein folding is of great importance not only in the field of medicine but also in biotechnology, due to the growing demand for the development of correctly shaped and biologically active proteins.

In the abstract, the stable structures that could be generated from a single polypeptide chain are innumerable. Often, the literature shows that the time required by a polypeptide long enough to "probe" any possible conformational state to find its native state would be comparable to the age of the universe. Once the denaturating agent has been removed, however, a denatured polypeptide employs a physiological time scale to spontaneously summarize the correct conformation. In light of these observations, studies on protein refolding focus on the intermediate folding states and the characterization of pathways that are followed during protein refolding. The occurrence of these intermediate states of protein folding has led to the hypothesis that folding takes place through defined pathways and through the cooperation of multiple interactions that progressively stabilize the state of a protein.

Ideally, the folding-free state of a protein is the so-called random coil, in which the possible conformations are many, even for a small protein. At this stage, a polypeptide should be found in a state where its chain would be as large as possible in space, and noncovalent interactions that would normally stabilize the native state would be nonexistent. Depending on the type of denaturating agent used, denatured proteins can show tidy local three-dimensional structures. For example, if denaturation is obtained under a pH that is similar to physiological pH or under mild temperature conditions, local interaction structures may be observed. This indicates that in order to maintain a protein in a completely denatured state, radical physical or chemical treatments are necessary. It is, therefore, permissible to suppose that the proteins in the cell, which are synthesized as linear polypeptides on the ribosome, begin folding during their own synthesis.

Under certain conditions, many proteins are found in a stable conformation that is not one of complete folding or even of complete steric disorganization. The occurrence, for example, of semisolid globular conformations (molten globules) is currently well documented. An important property of these structures is the presence of hydrophobic surfaces exposed to the external environment, which makes them susceptible to mutual aggregation and in which they transition to a state of complete cooperative-type folding. These observations suggest that the semisolid globular structure is a macromolecule, with secondary structures similar to those of the corresponding native protein but without stable tertiary structures. However, it should be noted that, although the number of possible conformations for the formation of a semisolid globular structure is still enormous, it is much smaller than that which is necessary for folding from a completely disorganized protein.

In vitro refolding is also believed to occur through several pathways involving one or more relatively stable folding intermediates. The process may start with a collapse of the hydrophobic regions within the molecule, with the formation of stable secondary structures that provide a framework for subsequent folding, or with the formation of interactions of weak bonds such as disulfide and hydrogen bridges, which stabilize polypeptides in specific conformational states. Each of these mechanisms can operate in concert during the initial stages of refolding. As a result, folding intermediates are formed that resemble the semisolid globular structures described above.

It is also believed that the conversion of the folding intermediates into the native state is a cooperative process in which the occurrence of an interaction makes the next interaction more stable. This "combined stability" will, therefore, be greater than that obtainable with the sum of the two interactions alone. Such a model implies the acquisition of the protein form by degrees, in a process in which a cooperative interaction is energetically favored compared to a noncooperative one. According to this hypothesis, the refolding of a denatured protein could occur through distinct and parallel pathways that involve the onset of different intermediate forms but that, by collaboration, lead to a well-defined native state.

In vitro experiments certainly have a significant value in defining the types of intramolecular interactions that drive the folding of proteins but hardly reflect the processes that take place in the folding of proteins within the cell. In fact, in vivo, the physiological temperature and the high local concentration of both total proteins and unstructured polypeptides will strongly favor the productive aggregation interactions toward a correct folding route.

A further difference between in vitro protein folding and in vivo protein folding is the enormous dissimilarity in the level of complexity of the environment perceived by proteins within cells. A further consideration is that, in the cell, proteins are synthesized by degrees on the ribosome. Since the concentration of the protein chains arising in the cell is high, these will be particularly prone to aggregation. Therefore, there must exist within the cell a mechanism by which such aggregations are favored or prevented, depending on the folding that must be achieved. Finally, one should recall that in the cell some slow steps in the folding of proteins are catalyzed by specific enzymes and emphasize the role of molecular chaperones in the folding process.

After this scientifically detailed discussion of the protein folding process, we now turn to more general and philosophical considerations that are not uniquely related to this specific process.

Philosophical Considerations on Protein Folding: System, Environment, and Self-Organization

Beginning first with the differences between in vitro folding and in vivo folding, we wish to highlight the philosophical meaning and significance of the highly complex cellular environment in which a process takes place within living systems. From a typically reductionist perspective, each process is studied separately from the others, and this view is useful for understanding the intrinsic characteristics of that process. But we must never forget that natural processes always occur together and influence each other. The systemic perspective is able to consider processes independently when needed and interconnectedly when necessary.

If we return to protein folding, a second philosophical consideration emerges, that is, the important role in determining the resulting and final stable structure played by the environment within which this process occurs. From the philosophical point of view, this introduces the problem of what an entity is when it is isolated and what it is when it is in interaction. This is a similar problem to that of isolated atoms versus atoms in the molecule, although in the molecular case the atoms join to form a new system while, in this macromolecular case, the nonisolation can lead to different forms of the same macromolecular entity.

We have established that the in vitro protein often has a stable configuration that is very similar to the native configuration (in vivo), although the latter depends both on the primary structure of the protein (molecular structure) and the environment in which it is located. The similarity between the two configurations may seem "strange," if the in vivo configuration depends on the environment and is obtained in a completely different manner (think only of the presence of ribosomes, enzymes, and molecular chaperones). Quantum mechanics can help us understand this strangeness. We have seen that in quantum mechanics the global states of the system are energetically separated. If this separation is large enough that the environmental interaction cannot considerably mix two isolated states, we can presume that the state assumed by the system in the environment is very similar to its isolated state.

A third philosophical consideration concerns the macromolecular form, which differs conceptually from the structure of the macromolecule. This is what leads us to say that macromolecules constitute a level of complexity greater than molecules because they have properties not explainable by or deducible from the molecular structure alone, as we saw in the denatured protein. However, once the macromolecule "works" within a cellular environment, its form is evolutionarily "selected" by adaptation to the environment and disease, and the different foldings of the protein are evidence of this adaptation.

These philosophical considerations open a general discussion on what should be understood as "environment," on the relationship "body/environment," on what is implied when we distinguish an aspect of reality in system + environment, as well as other philosophical and scientific issues of this type. The introduction to this book already discussed the essential philosophical and scientific concept of environment in the case of open systems and within our systemic perspective. Here, we wish to indicate that every systemic process, of which the folding of proteins is a good example, is driven both by intrinsic (in this case, macromolecular) and by extrinsic (environmental) aspects. Moreover, the relationship between intrinsic and extrinsic aspects, in this case as well as in others, is technically complex. Every system + environment, every ecosystem, is concomitantly both a generator and a receptor.

The same rules of interaction are shaped reciprocally, but this is not always highlighted in scientific theories. Edgar Morin tells us that the competitive character of natural selection, which Darwin's theory highlighted so well, is part of the competitive category of eco-organization, and selection integrates both competition and complementarity. The selection process thus enters a recursive ring with the integration, one in which the integration product retroactively selects what integrates it, while the selection product retroactively integrates what selects it. Therefore, evolution is not only the effect or product but also the cause and co-production factor of the complex conceptual ring. Let us move from the example of protein folding to another, more general aspect of what has been discussed thus far.

Statistics and Self-Organization

While always starting from the specific case considered, we wish to underscore a general inference underlying the statistical calculations applied to the folding process and many other cases within the molecular world. In our opinion, all these cases involve an error of "principle," an error that is linked to the lack of understanding of the role played by the chemical explanation of the material world. The problem of statistical analyses is "intrinsically" wrong, and this becomes quite obvious when examining the scientific literature. For example, this is revealed when thinking about the probability that complex molecules could be formed in the early stages of the origin of life. Therefore, let us now attempt to clarify this idea.

These types of probability calculations actually consider the molecules and macromolecules (or their parts) as unstructured units, like colored balls, for which you can "count" all the possible permutations of the parts. However, as we have noted and will continue to note throughout this book, molecules of all sizes, as well as individual molecular pieces, are not colored balls to be mixed in all possible ways with equal probability for each mixture. Rather, they are structured pieces of reality with specific characteristics and interactions that permit few, and sometimes only one, largely privileged interaction.

If we think that all cases are equally probable, we see that the number of possibilities grows exponentially as the number of parts increases. This often leads to the conclusion that, given such a high number of probabilities and such a short time frame for each single interaction, the age of the universe would not suffice to "explore" all possibilities. This is one reason why some people may be tempted to postulate intelligent design or some other type of external intervention as an explanation for why the universe displays the kind of structure and order that it does within the relatively short time frame of its existence. What we seek to highlight here is that, underlying these arguments is a premise that must be questioned, the premise that all of the possible states are equally probable.

Let us consider, as a first example, a polymer that in a particular environment is formed with a specific orientation of its monomers. In chemistry, such a polymer is called *stereospecific*. Again, we could easily assert that the number of steps that form the stereospecific polymer in a specific environment would have to be enormous, and, even if two monomers came into contact every millionth of a second, the time required for the synthesis of the polymer in its stereospecific structure would be much larger than the age of the universe. Here too, however, we are assuming that the system must explore all the possibilities because they are equally probable. But the actual process does not need this exploratory phase. The system, in fact, follows the path of minimum energy that leads, through the privileged interactions that are both internal to the structure and external to the environment, to the final form energetically superior in that context. Stereospecific polymers such as cellulose and natural rubber are formed under certain natural conditions, in spite of statistical calculations that would appear to "prove" the impossibility of such formations.

Let us consider a second example: the aperiodic polymer we have already discussed. Given twenty different molecules (e.g., twenty different amino acids), what is the probability that a polymer of a specific sequence will form? Typically, it is said that this probability is 20^N, with N representing the number of monomers, and that typically for a protein it is a number in the hundreds or thousands of units. Referring to the example used by Richard Dawkins in *The Blind Watchmaker*, let us assume $N = 146$, which is the number of monomers in one of the four hemoglobin chains. Since 20^{146} is a huge number, one may conclude that such an aperiodic polymer in a specific sequence cannot be formed randomly. Dawkins states:

> The number of possible ways of arranging 20 kinds of thing in chains 146 links long is an inconceivably large number, which Asimov calls the "hemoglobin number." It is easy to calculate, but impossible to visualize the answer. The first link in the 146-long chain could be any one of the 20 possible amino acids. The second link could also be any one of the 20, so the number of possible 2-link chains is 20 x 10, or 400. The number of possible 3-link chains is 20 x 20 x 20, or 8,000. The number of possible 146-link chains is 20 times itself 146 times. This is a staggeringly large number. A million is a 1 with 6 noughts after it. A billion, 1.000 million, is a 1 with 9 noughts after it. The number we seek, the "hemoglobin number", is (near enough) a 1 with 190 noughts after it! This is the chance against happening to hit upon hemoglobin by luck.[11]

Suppose we now begin with the amino acid A_1 and suppose that this molecule interacts preferentially with A_2: in a "natural" manner among all the twenty dimers A_1A_x, with x = 1–20, A_1A_2 will be formed. Let us state this more clearly: all possible dimers are formed, but if the system must be at the minimum possible energy, at the end (at equilibrium) we will have a prevalence of A_1A_2 dimers with a more or less narrow distribution of dimmers centered around them. The process can be iterated 144 more times, and we will "naturally" get a prevalence of $A_1A_2 \ldots A_i$ polymer, a specific polymer sequence of minimal energy and a distribution of less stable polymer sequences, centered around the most likely sequence, that is, the most energetically stable sequence. This shows

[11] Dawkins, *The Blind Watchmaker: Why the Evidence of Evolution Reveals a Universe Without Design*, 45.

that the prevalence of a sequence is obtained "naturally" but not through the system building upon the 20^N possible sequences and then "choosing" the most stable one.

In practice, the system follows a path that, at each step (each successive addition of a molecule), uses the constraint of "minimum energy." One might argue that, if this is the case, we should always have only one possible distribution of sequences that is centered on the most stable sequence and that, therefore, we are not able to explain the variety of sequences found in nature. In fact, this erroneous assumption is implicit in the statistical approach taken in the scientific literature. However, there is an additional conceptual error that we want to highlight when we look back on the process of protein folding.

We have seen that many steps of this process do not occur in series but in parallel. For example, hydrophobic processes all occur together, and, at the same time, they structure different parts of the macromolecule. Moreover, these processes are cooperative, and, thus, the occurrence of one increases the probability of the occurrence of the other. In practice, a macromolecule of 146 units can be obtained by adding one unit at a time and, thus, in 146 steps. But it can also be obtained by joining two pieces built in parallels of 83 units, which are in turn obtained by joining two pieces of 42 and 41 units, and so on. In this way, it is clear that the number of steps is greatly reduced. Obviously, this does not imply that we must always use pieces of equal or similar size, but in this case, we can see that

$$146 = (83+83) = (42+41)+(42+41) = [(21+21)+(20+21)]+[(21+21)$$
$$+(20+21)] = [(10+11)+(10+11)+(10+10)+(10+10)+(10+11)+(10+11)]+$$
$$\ldots = [(5+5)+(5+6)+(5+5)+(5+6)+(5+5)+(5+5)+(5+5)+(5+6)]+\ldots = \ldots.$$

With a small number of steps, we have obtained the macromolecule with 146 monomers. This is also a response to the objection regarding the uniqueness of the macromolecule obtained. The possible obtainable molecules are many, depending on the pieces that are joined. This mechanism of melding pieces also explains why, in DNA, for example, there are many "palindrome" pieces that can be read indifferently as to direction. In this case, this depends on the fusion of preexisting pieces in one direction or the other.

In practice, the variety is generated "historically" by the interactions that are possible in that particular spatial and temporal point of the process, by the complex dynamics of formation, and by fragmentation and regrouping of different pieces from the specific concentrations of the twenty molecules at each point, and so on. In practice, by putting together "case" and "necessity," as Monod would say, one can obtain the "historical variety" and many specific macromolecular systems, as our example demonstrates.

One might think that Dawkins' claims regarding hemoglobin are outdated, but they are not. Let us examine a similar example, used in Eva Jablonka and Marion J. Lamb's book *Evolution in Four Dimensions*. These authors state:

> We can think of a DNA strand as a linear sequence of units or modules (the nucleotides A, T, C, and G) in which each site in the sequence can be occupied by any one of the set of four nucleotides. A nucleotide at a given site can be replaced by any other without it affecting other nucleotides in the strand. This means that a huge number of sequences are possible. How large this number is depends on the length of the sequence, but even when it is not particularly long, the number of different possibilities is awesome. For example, with a sequence of only 100 units, made up of four different modules, 4^{100} different sequences are possible. This is a number that we cannot even imagine—it's more than the number of the atoms in the whole galaxy! And a stretch of 100 nucleotides would be just a tiny fragment of a DNA molecule.[12]

At this point, we should note another scientific characteristic related to self-assembly that is potentially interesting but that has not yet expressed all of its potential. We refer here to the self-catalytic reactions that have led to the development of a sub-branch of chemistry known as systems chemistry. In the case of RNA, Addy Pross states:

> So how does a replicating RNA molecule manage to make an exact copy of itself from a mix of the four nucleotide building blocks and in just the right sequence, when the number of possible sequences is so staggeringly large? The answer lies in the ability of the RNA molecule to act as a template. . . . Importantly, a lock and key type fit ensures that only the appropriate building block connects to any particular location on the RNA template so that the nucleotide sequence in the newly forming RNA chain is not arbitrary, but is specified by the original RNA strand. . . . It is only when the RNA molecule acting as a template is added to the mixture that the nucleotide building blocks line up along the RNA chain in the proper sequence, lock into position, and link up, thereby causing a replica of the RNA chain to be created.[13]

To understand the importance of these self-catalytic reactions, it is necessary to know that, although in normal catalytic reactions the rate of production proceeds linearly, in these self-catalytic reactions the rate of formation of the

---

[12] Jablonka and Lamb, *Evolution in Four Dimensions. Genetic, Epigenetic, Behavioral, and Symbolic Variation in the History of Life*, 55.

[13] Pross, *What Is Life? How Chemistry Becomes Biology*, 64–68.

product proceeds exponentially and this makes a huge difference. Pross provides a useful narrative to help us understand the enormous difference between these two processes:

> Let's explain the difference by recounting the classical legend of the Chinese emperor who was saved in battle by a peasant farmer. When the emperor asked the farmer how he could reward him, the farmer took out a standard chess board and asked that he be rewarded with a quantity of rice, and that the required quantity be established by a simple formula—placing a single grain of rice on the first square, two grains on the second square, four on the third, and so on, right through to the 64th square. The request sounded absurdly modest and the emperor was surprised that the peasant would be happy with such a small reward. After all, how much rice could be needed? Half a sack, a whole sack? But the truth is that the amount of rice needed to comply with the peasant's request is spectacularly large. Mathematically the total number of grains of rice placed on the board would be $2^{64}-1$. That works out at close to $2*10^{19}$ grains— that's a lot of rice; more than could be found in the emperor's cellars, as well as in all the world's Chinese restaurants, and, in fact, more than exists anywhere on the entire planet. That quantity of rice, if it existed, would cover the entire earth's surface to a depth of several centimeters. By comparison linear growth, as expressed by the catalytic path, would be the equivalent of placing a single grain of rice on each of the 64 squares. Hence the total amount of rice placed on the chess board would be just 64 grains![14]

Pross himself, however, reminds us that "systematic study in the area was only initiated some twenty-five years ago and many chemists remain unaware that such a field even exists."[15]

Finally, we point out that such a statistical discourse demonstrates that the evolution that created a complex organ such as the eye, for example, could not have taken place in a single step but involved many subsequent steps, from generation to generation. This example is well illustrated in the process of "cumulative selection" discussed by Richard Dawkins in *The Blind Watchmaker*. Here "cumulative" means that, once a variation has been created and has resulted in an improvement in the environment in which the organism lives, the variation is maintained in the next generation and, therefore, improvements accumulate with time. Important changes, such as eye development, can thus be achieved.

In this way, a small amount of order is added to the system at each step, and, when looking at the process between the initial state and the end state of the

[14] Ibid., 64.
[15] Ibid., 75–76.

sequence of steps, one can arrive from disorder to order without having to infer any planning or design. It is certainly true that by breaking a not very probable process into smaller, more probable processes that occur in series, the number of permutations that are possible for the overall process is greatly reduced. The folding of proteins, however, shows something more. On the one hand, the individual steps of the process take place in a parallel manner and not only in series. On the other hand, many of these individual steps are "cooperative," thus creating a "combined" effect that will be greater than the additive sum of the steps.

These examples are part of the broader problem of the self-organization of a complex system within a specific environment. We will return to this general problem, since it is fundamental to understanding the order that can be generated in living systems. At this point, however, we want to add another piece to this discussion with an extensive quotation from Moreno that further widens the field.

The spontaneous emergence of order could take two different forms: self-assembly (SA) and self-organization (SO). Unlike evolution, both SA and SO are widespread phenomena, which do not require very complex elements or systems. Self-assembly is a process in which a set of (randomly distributed) elements group together, forming a stable structure (an order), e.g., a crystal. This process is due to the material properties of the elements, namely, to the forces acting among them. Thermodynamically, self-assembly can be described as a process towards equilibrium, ending in a stable structure. In both SA and SO there is the emergence of order from a set of randomly interrelated elements. However, in the case of SO this order is not a consequence of the structure of the constitutive elements, but of certain boundary conditions in far-from-equilibrium (FFE) conditions: the emergent order is essentially dissipative. Given certain specific boundary conditions in FFE a set of local interactions becomes nonlinear, and a collective behavior—a macroscopic pattern—emerges. Now, the maintenance of this pattern is not only a consequence of the given boundary conditions, but also a result of the inherent recursivity of the process: once it appears, the pattern constraints the dynamics of the system components so that the produced pattern in turn produces itself. For instance, in the case of Bénard convection cells, the emergent pattern (the creation of hexagonal cells) contributes to the self-maintenance, because the fact of belonging to a certain cell is what makes a molecule turn to the left or the right. Thus, recursivity and removal from thermodynamic equilibrium is the key feature of this concept of organization. Though very different in nature, both SA and SO are important sources of order. In fact, many systems show both forms. However, only SO really holds the dynamical and functional senses of the idea of organization. The term "organization" implies not only order, but also the usefulness of this order that it effectively does something. And for this "doing", a continuous process is implicitly necessary. SO is therefore the ground of any organization as it is a dynamical form of order that contributes to the creation and *maintenance* of itself. The minimal (because self-sustaining) meaning of the terms *task* and *function* is that something is contributing to the maintenance of the very organization in which it appears. As we shall see, this internal sense of usefulness that appears in SO is the key for allowing a process of increase in complexity.[16]

Following these broad philosophical considerations that began from an important process like folding and have focused on many general aspects, let us now return to biological macromolecules, to the history of this concept, and to the scientific and philosophical difficulties that were overcome in order to "enrich" and "complexify" the study of living systems from a chemical point of view.

Determination of Different Protein Structures

The organization of macromolecules is complex. That DNA has a double helix structure and that proteins have more than one type of structure are today well-known facts, so we will not delve into the organization of macromolecules. Our aim is not to conduct a structural analysis of macromolecules but to highlight the general aspects implicit in that organization, as we have emphasized with regard to protein folding. As an example, we will discuss the determination of the primary and secondary structure of proteins, and, in a later chapter, we will discuss the structures of DNA.

Let us keep in mind some definitions. The primary structure of a protein refers to the sequence of amino acids that make up the specific protein. The secondary structure contains regions of amino acid chains that arise from the hydrogen bonds formed between atoms of the polypeptide backbone. The tertiary structure is that which is determined by the shape that the polypeptide chain can take at the conformational level, what we have called its native form. Its periodic character is linked to the regular character (e.g., helical or sheet) that individual pieces of macromolecules assume in space. The "tertiary structure" appears in proteins because the local "secondary structure" organizes itself spatially, creating a global regularity. Both the secondary and tertiary structures are stabilized by weak interactions (hydrogen and disulfide bonds but also hydrophobic, ionic, etc.) that connect specific groups of the sequence, sometimes even distant groups. The "quaternary structure" occurs when several polypeptide

---

[16] Moreno, "A Systemic Approach to the Origin of Biological Organization," 246–247.

chains (called protomers) are aggregated. Recently, our understanding of the organization of proteins has been further complicated with the addition of "supersecondary structure" and of "domains."

From a historical point of view, in order to determine the order of amino acids in a protein sequence, what we have defined as its primary structure, we began by using simple chemical methods. It was understood that fluoro-dinitrobenzene (FDNB) formed a bond by reacting with the free amino group at the end (N-terminus) of a polypeptide chain, and it was then possible to detach the terminal amino acid bound to dinitrobenzene. Using the later technique of chromatography, it was then possible to identify this specific amino acid.

In 1945, Frederick Sanger studied insulin using this method and found that two dinitrobenzene compounds were formed. One of these compounds was bound to glycine, and the other was bound to phenylalanine. This meant that there were two distinct polypeptide chains that started with these amino acids. The chain that began with glycine was named A, and the chain that began with phenylalanine was named B. These two chains were connected by a disulfide (sulphur-sulphur) bridge that Sanger was able to break chemically to obtain individual chains. By repeatedly using this method with FDNB, amino acid after amino acid, it was possible to determine the composition of a relatively short polypeptide. In order to obtain these polypeptides, the test protein was reacted with some enzymes that would selectively break it up. It was possible, by using this laborious method, to determine the order of the amino acids in a protein.

In 1951, Sanger was able to sequence the amino acids in the B-chain of insulin. It consisted of thirty amino acids that began with phenylalanine and valine, and ended with lysine and alanine. This was the first aperiodic polymer that was determined experimentally. In fact, before this analysis, the dominant hypothesis (the Bergmann–Niemann hypothesis) stated that there was some particular order of amino acids in proteins that could help clarify the mechanism of protein synthesis. Sanger stated that "[a]n examination of the structure B, however, fails to reveal any simple periodic arrangement of the residues, nor is it possible to formulate any general principles which might govern the order of amino-acids along the protein chains."[17] The polymer was, therefore, really aperiodic.

Two years after the clarification of chain B, the clarification of the sequence of amino acids in insulin chain A allowed scientists to determine the primary structure of this protein. In general, determining the structures of other proteins was a complex experimental and theoretical problem that involved many scientists and that involved X-ray spectroscopy as an experimental technique. We will not discuss this problem here in detail, but, to support this claim, we will focus on the

---

[17] Sanger and Tuppy, "The Amino Acid Sequence in the Phenylalanine Chain of Insulin. 2. The Investigation of Peptides from Enzymic Hydrolyzates."

determination of the structures of myoglobin and hemoglobin, which led to Max Ferdinand Perutz and John Kendrew being awarded the 1962 Nobel Prize.

Structure and Shape in Molecules and Macromolecules

To close this chapter in which we discussed the discovery of new levels of complexity in macromolecules, we must reexamine scientific and philosophical differences between the concepts of structure and molecular form and explain their role within both molecules and macromolecules. To provide more details, we will refer extensively to a recent paper on this topic co-written by Elena Ghibaudi, Luigi Cerruti, and Giovanni Villani, the title of which is "Structure, Shape, Topology: Entangled Concepts in Molecular Chemistry."[18]

As this chapter emphasizes, molecular structure is a systemic concept in that it transmits information about the internal organization of an entity (a molecule) whose constituents (atoms) merge to form a system. Chemists are aware that a molecule is a new entity distinct from its constituent atoms, with peculiar properties and a specific name. In addition, each possible organization of the same components bears a distinct name. A molecular structure provides information on the distribution of interatomic interactions. Here, we point out the nonequivalency of the terms *interaction* and *chemical bond*. The following are some implications of the systemic view of molecular structure:

(1) The systemic character of the molecule involves the global interaction between all the atoms that compose it. However, some of the atoms interact strongly, while others interact so weakly that they can be considered approximately noninteracting. The so-called structural formula encodes these preferential interactions as direct links between adjacent atoms. For example, the structural formula of a water molecule shows the oxygen atom bound to both hydrogen atoms, while the latter are not interconnected.

(2) A molecule is not simply an ordered aggregate of atoms. The atoms within the molecule are different entities from their initial condition. They are distinguished by isolated atoms and also by atoms of the same element within other molecules. Quantum mechanics (QM) highlights this condition of being part of a system through the assignment of fractional charges to each atom within the molecule. Here, we simply want to point out that atoms within the molecule must be considered in situ

---

[18] Ghibaudi, Cerruti, and Villani, "Structure, Shape, Topology: Entangled Concepts in Molecular Chemistry."

(i.e., in a particular relational context), as regards both their intrinsic and their relational properties. This means that atoms of the same element can be different, even if they are designated by the same symbol. They show an individual character that is correlated with their position within the molecule.

The idea that an atom is modified by becoming part of a molecule is a discovery of modern science that is far removed from the classical concept of atom. It is a consequence of the impressive conceptual evolution undergone by the concept of atom throughout modern scientific history. As previous chapters have discussed in detail, the atom conceived by the Greek philosophers was constitutionally immutable. Material changes were interpreted in terms of aggregation and separation of atoms that were conceived as rigidly invariant parts of matter. The idea of the immutability of the atom was initially challenged by Dalton's atomic theory and gradually changed during the 19th century. The structured atom of the 20th century, on the other hand, while it enters higher structures such as molecules, must first be dismantled and then restructured within the new system. The chemical reaction is precisely this process of deconstruction and restructuring.

(3) The molecular structure and the other structures of macromolecules (such as the tertiary structure of the proteins, which is called native) bear small perturbations. This resilience is crucial, as it marks a boundary between the molecule and its environment, e.g., it allows identification of chemical individuals with specific properties instead of a continuous energy landscape in which changes occur continuously. QM associates this resistance with the fact that the bound electronic states are quantified; that is, they are separated by a finite energy gap. Structural changes occur whenever the scale of the environmental disturbance allows the system to overcome the energy gap that separates one structure from the other.

From the molecular and/or macromolecular structure, it is deduced that the chemical individuals (molecule, macromolecule, and molecular pieces like polyatomic ions, etc.) are systems and that they must be treated according to the systemic perspective. This is the central idea that is leading us, in this book, from the atom to biological macromolecules, which are essential in living systems, and the "concept of structure" plays a fundamental role in this path. In simple cases, this concept is identified with that of organization. However, in the complexity of systems, the existence of multiple structures leads to the differentiation of these two concepts.

The concept of molecular shape is conceptually different from that of molecular structure, although these two concepts are often confused with each other, as the following examples will show. We choose to quote Richard Wolley on this issue for two reasons. First, his statement signals the importance of the concept of molecular structure, defining it as the "central dogma" of explanation in the molecular world, and, second, it highlights the classic confusion between molecular shape and molecular structure. The concept of *molecular shape* is widely used in the molecular sciences. However, providing a univocal definition of this concept is difficult. The problem is complicated by the widespread use of molecular shape as a synonym for molecular structure. The following extract is a clear example of such confusion. Wolley states:

> Those sciences that are concerned with the molecular aspects of the properties of matter, principally chemistry, but also molecular physics and biochemistry, are founded on the belief that all experiments involving molecules can be understood in terms of the relative dispositions of the constituent atoms in the molecules. This idea of *molecular structure* (or *"molecular shape"*) has been fundamental to the development of our understanding of the physicochemical properties of matter, and is now so familiar and deeply ingrained in our thinking that it is usually taken for granted—*it is the central dogma of molecular science.*[19]

The shape of a macroscopic object, identified by its external surface and characterized by a discrete number of points, can be reported in a three-dimensional graph. In the case of a molecule, its outer contour surface depends on the scale of size and energy at which it is studied and can only be identified by reference to specific models introduced to study a particular molecular property within a context. The concept of *molecular shape* is, therefore, dependent on the model and the property that this model wants to highlight. As a surface of separation between the molecule and the environment, the molecular shape strongly depends on both the molecule and the environment in which it is located.

After highlighting the substantial difference between the two concepts of shape and molecular structure, we point out that this difference can be important in the molecular world, but it is absolutely essential in the macromolecular one. In a molecule composed of two atoms, for example, these two concepts are in *biunivocal* relation because it can only be linear, regardless of the environment in which such a molecule is placed,. A molecule of three atoms, however, can be linear in one environment and angled in another. In this case, we are referring to

---

[19] Wolley, "Must a Molecule Have a Shape?"

a molecule in two conformations. But if the different forms are evidence of a different structure, we are referring to two different molecules.

As the number of atoms grows, it can occur that the same molecule (a structure) can have different stable conformations with different shapes. When we pass from a molecule to a macromolecule, such as a protein for example, the simple structure is determined by the succession of amino acids in the sequence and does not determine more important properties of the macromolecule. The difference between the properties of a macromolecule in its native form and the denatured molecule is a clear example of this. Therefore, to characterize the role of a macromolecule in a cellular environment, we must not only indicate its molecular structure but also its shape, and it is this additional information that characterizes the functional specificity of a macromolecule, as we have already mentioned and as we will discuss in detail in the next chapter.

What follows is a discussion of the increase of complexity in a proteome or protein field.[20] Contemporary protein science has been restructured by the discovery of the natural abundance of functional intrinsically disordered proteins (IDPs) and protein hybrids that contain both intrinsically disordered protein regions (IDPRs) and ordered regions. We can state that disorder-based functionality complements the functions of ordered proteins and domains. With their exceptional spatiotemporal heterogeneity and high conformational flexibility, IDPs/IDPRs represent complex systems that act at the edge of chaos and are specifically tunable by various means. IDPs are not homogeneous but represent a very complex mixture of a broad variety of partially foldable, potentially foldable, differently foldable, or completely unfoldable segments.

This behavior of an IDP as a highly frustrated system that does not possess a singular folded state is reflected in its free energy landscape, which is relatively flat and lacks a deep energy minimum that can be found within ordered proteins. Such a flattened energy landscape is extremely sensitive to different environmental changes that can modify the landscape in several different ways, lowering some energy minima while raising some energy barriers. This explains the conformational plasticity of an IDP/IDPR, its extreme sensitivity to changes in the environment, its ability to interact with multiple different partners, and, consequently, its ability to fold in different ways.

While moving from the classical structure–function paradigm, in which one gene encodes for a unique amino acid sequence that folds into a unique 3D structure responsible for the unique biological function of a given protein, one must pass to the structure–function quasi-continuum concept in which a single gene can produce multiple proteins with many functions. IDPs/IDPRs are not characterized by the reduced informational content of their amino

---

[20] Uversky, "Intrinsically Disordered Proteins and Their 'Mysterious' (Meta)Physics."

acid sequences, and their amino acid alphabet is not decreased in comparison with the alphabet utilized in the amino acid sequences of ordered domains and proteins. In any case, the (non)foldability of proteins is encoded in their amino acid sequences. Similarly to ordered (foldable) proteins whose unique biologically active structures are formed based on the information included within their amino acid sequences, the ability to not fold and still be functional in the absence of unique structures is also encoded in the specific features of the amino acid sequences of IDPs/IDPRs.

For extended IDPs, these features include the presence of multiple uncompensated charged groups (which are usually negative) that define the high net charges of many IDPs at neutral pH, combined with a low content of hydrophobic amino acid residues. This means that, from a physical viewpoint, the lack of compact structure in a protein with low hydrophobicity and high net charge makes perfect sense, since strong charge–charge repulsion in such a polypeptide is not compensated by strong enough hydrophobic attraction. Moreover, IDPs are known to gain a more ordered structure based on temperature, and they are usually more structured at higher temperatures, due to the enhancement of the hydrophobic interaction, while being more disordered at lower temperatures.

Overall, IDPs/IDPRs are complex systems with structurally and functionally sophisticated heterogeneous organization. Therefore, they define the possibility for a protein molecule to be multifunctional and to be involved in the regulation of, the interaction with, and the control from multiple structurally unrelated partners.

Instead of the classical, but heavily oversimplified, "one gene–one protein–one structure–one function" model, the actual protein structure–function relationship is described by the more convoluted "one-gene–many proteins–many functions" model. This confirms the principle that the complexity of a biological system is primarily determined not by the genome size, but by its proteome size. The number of functionally different proteins is known to dramatically exceed the number of protein-encoding genes (e.g., the human genome is approaching 20,700 genes, but the actual number of functionally different proteins is in the range of a few million).

Over the years, the process of understanding protein interaction has undergone a dramatic shift from a highly static to a very dynamic view. It is clear that internal dynamics is crucial for the biological activity of many (if not all) proteins, indicating that protein functionality requires at least some degree of conformational flexibility and structural dynamics. Even in ordered proteins, functional dynamics involve the movements of not only individual amino acid residues or groups of amino acids in an active site relative to each, but also displacements of entire domains. These function-related movements that occur in a wide range of time scales, from femtoseconds to seconds, are required for the facilitation of catalytic activity. Despite their lack of stable structure, many IDPs/IDPRs are promiscuous binders that are never nude, being always-complexed, and invariably interacting with various partners via multiple binding scenarios. We can say: "we need more chemistry, not less" to study these systems.

Since IDPs/IDPRs are system complexes at the "edge of the chaos," they are characterized by emergent behavior that is based on intricate self-organization processes, leading to the appearance of unanticipated novel structures, patterns, and properties. In spite of being complex biological systems that are seemingly positioned at the edge of chaos, IDPs are not completely random entities, since they have evolved some adjustable, controllable, regulable, tunable, and, often, very specific properties required for their biological functions. In conclusion, IDPs/IDPRs are very different from ordered proteins and domains at multiple levels. But these proteins and regions also have recognizable amino acid sequences with multiple biases, which makes their behavior easily predictable.

12

From the Gene to Metagenomics: The Frontiers of Molecular Biology

In the context of the study of living matter, the term *molecular biology* has a very precise meaning and is often linked to the concept of the gene at the molecular level of DNA. In the literal sense, however, "molecular biology" is much broader and constitutes one of the fundamental topics of this book. Thus, although molecular biology is explicitly discussed only in the later chapters, it also tacitly informs the early chapters. Since molecular biology involves molecularizing the study of living beings, we can trace the historical and epistemological path of this concept, in its broad sense, at least up to Lavoisier's analysis of respiration.

In this chapter, however, we are concerned only with the narrow sense of molecular biology and with its development. To this end, we will begin with Gregor Mendel's research and his theory of particulate inheritance, which stipulates that it is through units of inheritance that living beings transmit to their progeny the information relative to their specific morphological and physiological characteristics. Starting from experimental results obtained by repeated crossings of pea plants, Mendel formulated the laws of the transmission of characteristics from one generation to the next. This is a fundamental turning point since Mendel's ideas were also fundamental to the eventual molecularization of biology.

After the rediscovery of Mendel's work in the early 20th century, biologists clarified the molecular vector of biological inheritance; that is, they identified and determined the structure of the molecule that stores and passes on information. After the development of the concept of "gene," this area of research acquired the name genetics. The information that is stored in the DNA molecule is closely related to its molecular structure. Thus, we will now trace the historical and philosophical development, beginning with the unit of heredity, leading to the concept of the gene and its molecularization in DNA, and culminating with the latest developments in epigenetics and metagenomics.

This is the only chapter in the book that is dedicated to a distinct molecule or, more specifically, to a distinct class of molecules. Along with the water molecule, DNA is undoubtedly the most important of all molecules. However, while H_2O is a small molecule formed by three atoms, DNA is a macromolecule containing billions of atoms. DNA is not a chemical formula but an acronym that stands for deoxyribonucleic acid. The acronym DNA is not only scientifically important

From the Atom to Living Systems. Marina Paola Banchetti-Robino and Giovanni Villani, Oxford University Press.
© Oxford University Press 2023. DOI: 10.1093/oso/9780197598900.003.0013

but has acquired a significance that transcends science itself. This molecule is the seat not only of the identity of all living beings but also of all biological species.

This acronym has even infiltrated vernacular expressions, such as "it is in this person's DNA," an expression that links a person's psychological traits to their biologically inherited characteristics and that begs the question in favor of nature, as opposed to culture, as the source of one's psychology and behavior. Moreover, DNA is used to differentiate living species, and, more specifically, to trace the "transformation" of the human species from its original genus *Homo* to modern *Homo Sapiens Sapiens*. Finally, the ability to manipulate DNA renders possible the selective alteration and "improvement" of the human genome, thus giving rise to a great deal of philosophical debate and a wide range of ethical concerns about genetics.

Mendel: An Innovator Rediscovered

It was the Abbot Gregor J. Mendel, a contemporary of Charles Darwin, who first established the laws of character inheritance in a memoir entitled *Versuche über Pfanzen-Hybriden* (*Experiments on Plant Hybridization*). Between 1857 and 1868, Mendel devoted himself to a series of experiments on pea plants (*Pisum sativum*) in the garden of his convent. Analyzing the results of the crossings carried out on these plants, he articulated the three laws on the transmission of hereditary characteristics that have now acquired his name:

1. The Law of Dominance and Uniformity (Mendel's First Law) stipulates that from the intersection of two pure line individuals that differ by only one possible characteristic in two forms, a first generation of hybrid individuals is obtained in which only one of the two alternative forms of the characteristic is manifested. The manifested characteristic is defined as dominant, while the other characteristic is defined as recessive.
2. The Law of Segregation (Mendel's Second Law) stipulates that, from the crossing of first-generation hybrids, second-generation hybrids are obtained in which both forms of the characteristic are manifested. The dominant and recessive forms are in the ratio of 3:1.
3. The Law of Independent Assortment (Mendel's Third Law) stipulates that, by crossing individuals that differ by two or more characteristics, the latter are transmitted independently.

From these laws, it is concluded that each hereditary characteristic has a "unit of information" that can present itself in several forms. Crossing two "pure" individuals (i.e., two individuals who give only one form of a characteristic for

194 FROM THE ATOM TO LIVING SYSTEMS

several generations) will only express the dominant characteristic in the first generation if there are only two variants of such unity in which one is the dominant variant and the other is the recessive variant. Since the recessive characteristic has not "disappeared" in this generation but is simply not expressed, later hybrid generations will express the two forms of the characteristic according to the proportions given by simple statistical laws, taking into account that the recessive form of the characteristic can be expressed only in the presence of another recessive form and so only when the hybrid individual inherits two recessive forms of the same characteristic. As Richard Dawkins tells us in *The Blind Watchmaker*:

> Mendel showed that we don't blend our inheritance from our two parents. We receive our inheritance in discrete particles. As far as each particle is concerned, we either inherit it or we don't. Actually, as R. A. Fisher, one of the founding fathers of what is now called neo-Darwinism, has pointed out, this fact of particulate inheritance has always been staring us in the face, every time we think about sex. We inherit attributes from a male and a female parent, but each of us is either a male or a female, not hermaphrodite.[1]

Unfortunately, Mendel's work was forgotten for decades and was rediscovered only at the beginning of the 20th century by the English biologist William Bateson, at which point Mendel's theory of particulate theory was finally merged with the concept of the gene.

The Concept of the Gene

The concept of the gene falls between that of the macromolecule of DNA and that of the nonmaterial notion of inherited characteristics as described by Mendel. When, in the 1890s, the Dutch botanist Hugo de Vries came across the concept of gene, he immediately realized that this idea would revolutionize our entire vision of nature. This concept was akin to the physicists' and chemists' concepts of atoms and of molecules in the sense that it was considered a basic unit. In the context of genetics, however, it was the basic unit of inheritance:

> It will then be seen that the character of each individual species is composed of numerous hereditary qualities, of which by far the most occur in almost innumerable other species. . . . For the building up of the sum total of all organisms, there is required a rather small number of individual hereditary characters in

---

[1] Dawkins, *The Blind Watchmaker: Why the Evidence of Evolution Reveals a Universe without Design*, 113.

proportion to the number of species. Regarded in this way, each species appears to us as a very complex picture, whereas the whole organic world is the result of innumerable different combinations and permutations of relatively few factors. These factors are the units which the science of heredity has to investigate. Just as physics and chemistry go back to molecules and atoms, the biological sciences have to penetrate these units in order to explain, by means of their combinations, the phenomena of the living world.[2]

Lamarck was convinced that, in order to adapt to their environment, animals strengthened or weakened certain characteristics, depending on when they were used. This effect of adaptation to the environment was transmitted to the off-spring, who would then be preadapted to the environment. Thus, Lamarckism explained "progress" in the animal world in a simple way: Animals gradually adapted to their environment and, in so doing, perfected themselves.

Darwin himself had made it clear that a theory of heredity was fundamental to his and to any other theory of evolution. For evolution to function successfully, inheritance has to meet both requirements of constancy and variance, as well as those of stability and mutation. Darwin imagined that the organs of all living beings produced tiny particles that he called gemmules, which contained hereditary information, and he called this theory pangenesis (i.e., "genesis of the whole"). In a letter to Asa Gray, he stated: "The chapter on what I call Pangenesis will be called a mad dream, ... but at the bottom of my own mind, I think it contains a great truth."[3] It is interesting to note here that the idea of "gemmule" seems to be a modern and scientific version of the medieval and Renaissance concept of *semina rerum* (or *logoi spermatikoi*), which was discussed in Chapter 1 and which points to the notion of "seeds" or particles that contain a *logos*—that is, something akin to information.

The main criticism against Darwin's notion of pangenesis was that, if the hereditary characteristics continued to "blend" with each other at each generation, what would prevent a variation from being immediately attenuated by this crossing and, thus, from disappearing with time? If the heredity had not been able to maintain the variation, to "fix" the altered trait, all the variations of the characteristics would have eventually weakened and then vanished. The mixture could not work. It was necessary for there to be atomic units of information, discrete particles that remained unchanged in the offspring, thus transmitting the variation from the parent to the child.

Mendel himself wondered what would happen if he crossed a tall plant with a short one. Would a plant of intermediate height be born? From his experiments,

[2] Vries, *Intracellular Pangenesis: Including a Paper on Fertilization and Hybridization*, 12–13.
[3] Darwin, "Charles Darwin to Asa Gray," in *The Life and Letters of Charles Darwin: Including Autobiographical Chapter*, 256.

Mendel discovered that single inherited charactèristics, such as long or short stems or green or yellow seeds, never mix. The hybrid characteristic, Mendel wrote, was not intermediate but resembled one of the parental forms. He thus defined as dominant the characteristics that manifested and as recessive those that disappeared. But what had happened to the recessive characteristic? Has it been consumed or eliminated? Mendel crossed short-to-tall hybrids with short-to-tall hybrids and discovered that, in the next generation, the recessive characteristic reappeared. Mendel did not use the term "unit of heredity," but he had discovered the fundamental characteristics of the gene.

In 1883, the embryologist August Weismann disagreed both with Darwin, whose idea of pangenesis stipulated that the information particles from the organism were collected and ordered in gametes, and with Lamarck, who believed that characteristics acquired with use were transmitted to the offspring. Instead, Weismann proposed that the hereditary characteristics were contained exclusively in the gametes, in what he called the "germ plasma." It was Hugo de Vries who then inquired about the material nature of the "germ plasma" and about whether the information contained within it was discrete. De Vries had not read Mendel, but he rediscovered the discrete nature of the transmission of characteristics through particles that he called "pangen." Although de Vries had dismantled the theory of Darwinian pangenesis, the choice of the term *pangen* does recall Darwin's term. The Danish botanist Wilhelm Ludvig Johannsen who later would shorten the term *pangen* to *gene*.

De Vries also wondered how the different varieties of a plant were born. He called these varieties "mutants," from the Latin *mutate*, which means "change," and this name met with great approval in biology. Mutations, for de Vries, originated spontaneously within a normal population and resulted in varieties in nature. These variants were hereditary and were transmitted in a discrete form to subsequent generations. Darwin's natural selection, according to de Vries, acted upon the variants so that they functioned as the "engine" of evolution.

The English biologist William Bateson was responsible for the positive reevaluation of Mendel's writings and discoveries, for coining the term *genetics* (from the Greek *ghénos*, meaning birth), and for defining this field as the study of inheritance and variability. According to Bateson, the social aspects of genetics were clear.[4] If genes were "information particles" that were independent of each other, then they could be isolated and individually selected, at which point "humanity will begin to interfere" with nature. Until this time, the name "gene" had not corresponded to a material substrate or a chemical species, nor were its workings or its location within the organism or the cell known. The term had been created to designate a function and, thus, it remained as an abstraction. It

[4] Mukherjee, *The Gene. An Intimate History*, 67.

was defined by what it did; that is, it was considered to be the vector of hereditary information. As such, it only expressed the fact that many features of the organism are specified in unique, distinct, and, therefore, independent ways.

In the 1890s, while conducting research on sea urchins at the Naples aquarium, the German embryologist Theodor Boveri proposed the hypothesis that genes resided in chromosomes. These were filaments that, after being dyed blue with aniline, were found to be located in the nucleus of the cells. After locating the genes carrying the genus in a single chromosome, Boveri's pupil, the biologist Nettie Stevens, concluded that all genes resided in chromosomes. The American biologist Thomas Morgan initially opposed Mendel's inheritance theory because he said that it was impossible for complex embryological information to be stored in discrete units within the cell.

After these units were "found" in the chromosomes, Morgan himself raised the following questions in his Nobel Lecture of 1934: What is the nature of the elements of heredity that Mendel postulated as purely theoretical units? What are genes? Now that we have located genes in the chromosomes, are we justified in regarding them as material units, as chemical bodies of a higher order than molecules? Mendel had discovered that each characteristic was independently inherited, while Morgan discovered that some genes behaved as if they were "associated" with each other. The gene responsible for the black color of *Drosophila* fly was associated with the gene that specified the shape of the wings. To Morgan, this meant that these genes were physically connected and were on the same chromosome. The chromosome was like a "thread" along which the genes were arranged. The gene, therefore, was not a "purely theoretical unit" but was something "physical" that was in a particular position on the chromosomes.

There was, however, an exception to the association of genes. Sometimes, a gene would dissociate from other genes and change place by passing to the homologous chromosome. Morgan called this phenomenon "crossing over." Over time, it was understood that genetic information can be mixed, coupled, and exchanged, not only between homologous chromosomes but also between different chromosomes and, finally, between organisms of different species. Morgan assumed that genes that never crossed over were both paired and close to each other in the chromosome. Those that could cross over were associated but also more distant on the chromosomes.

In his 1934 Nobel Lecture, Morgan stated: "Today we arrange the genes in a chart or map. The numbers attached express the distance of each gene from some arbitrary point taken as zero. These numbers make it possible to foretell how any new character that may appear will be inherited with respect to all other characters, as soon as its crossing-over value with respect to any other two characters is determined."[5] The chromosome was, therefore, like a necklace, and the genes were like pearls along the wire. But clarification was still needed regarding what these pearls were made of and what was the chemical nature of the thread of the necklace.

The Materialization of the Gene: DNA

Biologists had not yet understood the material composition of the gene because they had never encountered a free gene that would permit experimentation to determine its chemical nature. When a cell divides, the genetic material divides within it and is distributed between the daughter cells. Throughout the process, genes remain biologically visible but chemically impenetrable, locked inside the cell core. Sometimes, however, the genetic material does not pass from the parent cell to the daughter cell but is transmitted between two unrelated cells. This process, called transformation, was discovered by the English bacteriologist Frederick Griffith, who identified it in the *streptococcus pneumoniae* bacterium.

There were two strains of the *streptococcus* bacterium: a smooth strain (virulent) and a rough strain (relatively harmless). In an experiment, Griffith killed a sample of the smooth strain using heat and then injected it into mice. As expected, these mice did not get sick because the bacteria were dead. When Griffith instead injected the heated material along with a sample of the rough strain, the mice became ill. The virulence, which was linked to the capsule that coated the smooth strain, had somehow passed to the rough strain, thus rendering them smooth. In practice, the rough strain had synthesized the capsule from the information provided by the dead smooth strain. In addition, once the formerly rough strain were transformed into smooth strain, they maintained this characteristic in subsequent generations.

The obvious question to be answered was, "How could the chemical residues of bacteria killed by heat, basically a mixture of bacterial chemicals, have transmitted a genetic characteristic to a live bacterium?" With this experiment, Griffith had discovered a chemical that could exist outside the cell, like a messenger that carried genetic information from one organism to another. This "transformation" process rarely occurs in complex organisms such as mammals. But much simpler organisms such as bacteria can exchange genes. Caught in the passage between two organisms, the "carrier gene" exists as a pure chemical substance and can be analyzed by chemical means.

---

[5] Morgan, "The Relation of Genetics to Physiology and Medicine: Nobel Lecture (June 4, 1934," 315.

In his 1944 book, *What Is Life? The Physical Aspect of the Living Cell. Mind and Matter*, physicist Erwin Schrödinger boldly attempted to describe the molecular nature of the gene based on purely theoretical principles,. Given this book's resonance and influence, we will conduct a thorough discussion of it in a later chapter. At this point, however, we are primarily interested in Schrödinger's observation that the gene had to be composed of a particular type of chemical, that is, of a molecule that could display richly contradictory characteristics.

This molecule surely had to display some regularity. Otherwise, the routine processes of copying and transmitting information would not succeed. However, it must also be capable of irregularities. Otherwise, the immense variety of living organisms could not be explained. The molecule must be able to carry a huge amount of data, but it must also be able to be compacted into the cells. Schrödinger imagined that this molecule would have multiple chemical bonds arranged along the "chromosomal fibre" and that the sequence of the bonds would organize "the elaborate code that concerns all the future development of the organism." Order and diversity, message and matter. Schrödinger seems to have already pictured the DNA molecule in his mind.

By the 1940s, biochemists knew that chromatin, the biological structure in which genes reside, is composed of two types of chemicals: proteins and nucleic acids. The chemical structure of these two components was not then known, but quite a bit was already known about proteins. Proteins were very versatile and, therefore, were the best candidates to function as vectors of genetic information. It was known that proteins played an essential role in cell function and that, in order to live, cells needed many different chemical reactions, which were "activated and controlled" by proteins. Proteins also played many other cellular roles. They represented the structural components of the cell and allowed chemical "communication" between cells. Their role was, therefore, essential in the cell.

Nucleic acids, on the other hand, seemed to be secondary molecules. They were discovered in 1869 by the Swiss biochemist Friedrich Miescher, and, in the 1920s, two types of nucleic acid, DNA and RNA, were identified. Both types were formed by four nitrogenous bases—adenine, guanine, cytosine and thymine in the case of DNA and adenine, guanine, cytosine and uracil in the case of RNA. With only four bases, compared to the twenty amino acids proposed by Schrödinger, nucleic acids did not seem to resemble the unique chemical that he had imagined.

Working in the 1940s, using culture media with the addition of different enzymes, Oswald Avery determined that Griffith's "transformation principle" could not be a sugar, a lipid, or a protein. Instead, using an enzyme that degraded DNA, he understood that the transformation principle was precisely the DNA molecule. With various experiments involving ultraviolet radiation, chemical analysis, and electrophoresis, Avery confirmed the role of DNA, and, in 1943, he wrote to his brother, Roy: "If we are right—and of course that's not yet proven—then nucleic acids are not merely structurally important but functionally active substances. . . . that induce predictable and hereditary changes in cell."[6]

What was still not understood, however, was what enabled DNA to perform its function. To resolve this problem, a large number of scientists in the 1950s began to focus their studies on the structure of DNA. Among these many researchers, the names Francis Crick, James Watson, and Maurice Wilkins stand out and they were, in fact, awarded the 1962 Nobel Prize for their discovery of the double helix structure of DNA. Other noteworthy scientists are Rosalind Franklin and Linus Pauling, who also actively participated in resolving this problem. In this regard, we can mention Rosalind Franklin and R. G. Gosling's experiment using X-ray diffraction patterns to identify the structure of DNA. The story of this exciting discovery is known both in its "classic" iteration, which recognizes Watson and Crick as having discovered the structure of DNA, and in its "modern" iteration, which focuses on the "implications" of this discovery.

Without retracing all of this background, we will simply mention that the significance of this discovery is borne out by the fact that, on April 25, 1953, three consecutive articles[7,8,9] were published on pages 737–741 of volume 171 of *Nature*, which definitively answered the question regarding the structure of DNA. We now know that DNA consists of two long polymers that run in opposite directions and that form the regular structure of the famous double helix. The monomers of DNA are called nucleotides. Each nucleotide contains a phosphate group, a sugar, and a nitrogenous base. The four types of nitrogenous DNA bases are adenine (A), thymine (T), guanine (G), and cytosine (C). The nucleotides are attached to each other to form two long filaments, held together by hydrogen bridges (formed between the base pairs A-T and C-G and vice versa). There are about 3 billion base pairs in human DNA, and it is the order of these base pairs that determines the information carried by the DNA, which forms the "genetic code."

After the DNA structure was discovered, it became clear that each mutant organism lacks a specific metabolic function and that this corresponds to the activity of a single protein enzyme. Since individual mutants are defective to a single gene, the normal gene must carry information to fabricate the actor's metabolic function. In practice, the gene acts by coding information to fabricate a protein and the protein performs functions in the body. Unlike the DNA chain

[6] Avery, as cited in Siddhartha Mukherjee, *The Gene. An Intimate History*, 139.
[7] Watson and Crick, "Molecular Structure of Nucleic Acids: A Structure for Deoxyribose Nucleic Acid," 737–738.
[8] Wilkins, Stokes, and Wilson, "Molecular Structure of Nucleic Acids: Molecular Structure of Deoxypentose Nucleic Acids," 738–740.
[9] Franklin and Gosling, "Molecular Configuration in Sodium Thymonucleate", 740–741.

that exists mostly in the double-helix form, a protein chain can fold into space and take a specific shape, through a process that is now known as protein folding. The ability to acquire a specific configuration (native form) is related to the function that each specific protein performs. However, the question regarding how a DNA sequence could carry instructions to make a protein still remained unanswered.

In the 1950s, two French geneticists, Jacques Monod and François Jacob, inferred that an intermediate molecule, or messenger molecule, was needed to translate DNA into proteins. This messenger had to travel between the cell nucleus, in which the genes were located, and the cytoplasm, in which the proteins were synthesized. Working with Jacob, Sydney Brenner isolated the messenger molecule within some bacterial cells, and it was in this particular type of RNA that the DNA message was transcribed and then translated into proteins. Crick referred to the path or flow of information that always moved in the direction of DNA →RNA →proteins as the "central dogma" of molecular biology. The word "central" indicated the universality of this path. Crick suggested that, in some specific cases, it was possible for the flow of information to be reversed, that is, to move from RNA to DNA. But it was the work of Howard Temin and David Baltimore with retroviruses that led to the discovery, in the 1970s, of "reverse transcriptase," in which a single-strand DNA (cDNA) is synthesized based on the RNA sequence.

A key piece of the puzzle was still missing, however. Through a series of ingenious experiments, Crick and Brenner had understood that the genetic code must occur in the form of "triplets," that is, three DNA bases, and that each triplet specified only one amino acid of a protein. The triplet hypothesis was also supported by mathematics. With two sets of four bases, one could specify only 16 (4 x 4) amino acids. But proteins contain 20 amino acids, in addition to some triplets necessary to indicate the so-called stop codon, which indicates the end of a specific protein. At least one triplet of DNA bases was needed because, with the four DNA bases, there are 64 (4 x 4 x 4) triplets to specify a single amino acid.

The "Complexification" of DNA

Two principal modifications have been made to the description of DNA since this molecule was initially discovered. The first of these changes has been to overcome the idea that only DNA serves the function of protein coding. We explain this in more detail. Eukaryotes are organisms in which the genetic material is DNA in the form of chromosomes contained within a distinct nucleus, and these include all living organisms other than eubacteria and archaebacteria. In eukaryotes, the DNA sequence (called exons) that constitutes a gene and that can, therefore, be associated with the synthesis of a protein is noncontiguous and contains within it other nucleotide sequences (called introns), which are not used to encode the amino acids to be inserted into the protein.

Moreover, as discussed in the previous chapter, rather than using the classical but heavily oversimplified, "one gene–one protein" model, the actual DNA–protein relationship is best described by the more convoluted "one gene–many proteins" model. This is because, when introns are removed, exons can be aggregated in different ways. The gene is, therefore, formed by "exons" (the nucleotide sequences that encode the amino acids) interspersed with "introns" (noncoding sequences). Thus, the gene is not a continuous sequence of DNA nucleotides.

The formation of messenger RNA is, therefore, more complex because it is transcribed into a pre-RNA, which contains both exons and introns. Before passing into the cytoplasm, this pre-RNA is modified, so that the introns are cut off and the exons are joined together. This process allows one to obtain the sequence of nucleotides with the exact information for the insertion of all the amino acids that constitute the protein. In practice, only a small portion of eukaryote DNA (1.5%–2.0%) is used for protein synthesis. The rest consists of noncoding sequences, whose biological role is not yet fully known.

In the past, the noncoding part of DNA received the collective name of "junk DNA," a term first used by Susumo Ohno at the 1972 Brookhaven Symposium on Biology.[10] It is clear today that this name is not appropriate, if only because it would be unthinkable to define the vast majority of DNA as "junk." We now know that the so-called "junk DNA" actually contains sequences that perform essential biological functions such as regulating the activation of genes, as well as other sequences whose biological function is not yet understood.

The second modification to our understanding of DNA resulted from an observation by Jacques Monod, after he discovered that the growth of the *Escherichia coli* bacterium in the presence of glucose and lactose was "discontinuous." These bacteria activated the selective consumption first of glucose and then of lactose. Monod, in conjunction with François Jacob and Arthur Pardee, reflected on the nature of cellular metabolism, on how it was possible for the enzymes involved in this process to appear and disappear within a cell, and how enzymes were encoded by DNA just like proteins. They eventually hypothesized that genes could be regulated by metabolic inputs.

When a gene is activated or deactivated, the copy of the DNA coding protein is always stored in the cell and the metabolism acts exclusively on the amount of RNA produced. During glucose metabolism, the messenger RNA essential to produce enzymes that act on this sugar was abundant and the messenger RNA necessary to produce enzymes that act on lactose was suppressed. The

---

[10]   Ohno, "So Much 'Junk DNA' in Our Genome."

opposite occurred when lactose metabolism was predominant. The production of messenger RNA was coordinated, and a whole set of genes was activated or deactivated by the environment. An entire functional circuit, which Monod and his associates called the *operon*, could be switched either on or off.

In short, a group of genes possessed not only the information to encode a protein, but also the information to decide if and when to encode it. All of this data was contained in DNA, but it was also related to environmental inputs via proteins. In 1910, Morgan "associated" the genes with each other, and it finally became clear that a group of genes could act together because they were acting on the same metabolic pathway. It was also understood that the passage DNA →RNA →proteins was actually a cycle, DNA →RNA →proteins →DNA, that was written in chemical language.

This discovery also addressed the general problem of how an organism can have a fixed genome and yet respond in a prompt manner to a variable environment. It also explained how different cells could originate from the individual genetic heritage of the fertilized cell. The differences depended on the fact that in each type of cell a different subset of genes was "read" and "translated" into its protein products. This "choice" was determined by the chemicals already present in the cell, which in turn were contingent upon previous gene readings in that cell and upon the "chemical situation" in neighboring cells.

The entire embryo development process could be considered as a collective undertaking, and it was through this mechanism of gene regulation that cells could perform their functions not only in space but also in time. Moreover, since the DNA duplicating function could not be autonomous and had to be controlled by a protein for its replication, this process also had to be a cycle, DNA →RNA →protein →DNA, one that was activated or deactivated by precise "external" environmental signals. In 1958, the American biochemist Arthur Kornberg isolated the "DNA polymerase," the enzyme that is manufactured in varying amounts by DNA, under the inputs of age, nutritional status, and other environmental factors. This process is also cyclical, that is, the protein produces the DNA, which, in turn, produces the protein. Finally, it was eventually discovered that even the fundamental embryological problem of the passage from the single fertilized egg cell to a multicellular organism with its shape and structure is regulated in the same manner. An organism creates its own specific anatomy and physiology by mixing and combining hierarchies, gradients, genes, and proteins and by inserting all these in interaction within the environment.

It bears emphasizing that all of these processes are cyclical. All living beings have a surprising uniformity at the level of their molecular components, and this fact is evident especially in the genetic code. This genetic dictionary comprises sixty-four "words" of DNA, which are codons of three letters each. Each of these codons has a precise translation into the language of proteins. It is a fact of considerable importance that all living beings, however different they may appear to be on the outside, speak the same internal chemical language that is focused on the DNA →RNA →protein →DNA cycle. As well, the pivotal process of DNA in evolutionary replication is also cyclical because it exerts an influence on the likelihood of its own replication through organisms.

The cyclical nature of these biological processes is fundamental, and we have seen it at work also in other biological processes. Here, we wish to emphasize both its scientific and philosophical importance. As discussed in the introduction to this book, a cyclical explanation lays bare one of the greatest limits of reductionism. As already mentioned, there are at least two types of reductionism: Laplacian reductionism and hierarchical reductionism. In Laplacian reductionism, every single complex event is explained by determining the entities and the laws of interaction within the initial link of the chain, that is, the level of elementary particles. Once this is known, it is believed that each event at each higher level is, at least theoretically, implicit in the initial level. Hierarchical reductionism, on the other hand, considers that the whole of a complex system can be described with a hierarchy of organizations, each of which is only described in terms of objects that are one level down in that hierarchy.

In spite of their differences, both of these variants of reductionism imply that explanation can only take one direction, that is, bottom-up. However, a cyclical explanation such as the biological one that we have discussed involves the presence of both bottom-up and top-down explanations, with their explanatory closure in a cycle. Thus, from a philosophical perspective, the cyclical nature of the DNA process that we have described cannot be satisfactorily understood within a reductionist standpoint.

DNA as an Identity Factor of Living Species

So far we have discussed the role of DNA in a cell and, therefore, in a living organism. However, DNA plays another essential role, that of giving each living species its own identity. This role is specific and is based on the similar but not equal concept of macromolecules. The individuals within a species never have the same identical DNA, apart from the case of homozygous twins. Nevertheless, each species has its own unique DNA.

From the chemical point of view, two molecules are the same if they have the same structure. However, even in this case, they are not identical because they can have a different kinetic energy (speed), among other things. For this reason, within a pure sample formed by a single type of molecule and a given temperature, the molecules are distributed according to their kinetic energy. This implies molecules with a different ability to react and, therefore, "different" molecules.

Thus, it is problematic to identify the molecular structure as representing the specific identity of a molecule. When we move from a molecule with a few atoms to a macromolecule, we see that the simple molecular structure is not enough to identify the macromolecule. Hermann Staudinger, the father of macromolecular chemistry, states that "[i]n a simple and homogeneous substance, the molecules are all of the same size: the high-molecular compounds, on the contrary, consist mainly of a mixture of molecules of similar constitution, but of different sizes."[11]

This characteristic is not unique to biological macromolecules. Although we have just discussed the case of DNA in individuals of the same species, this characteristic is also found in inanimate synthetic polymers. In the latter case, for a pure sample, we speak of average molecular weight, which means that the specific polymer name that is used (e.g., polyethylene) is not constituted by a fixed number of monomers but by polymer chains of different lengths, from macromolecules that have different molecular structure.

From a philosophical point of view, this implies that two macromolecules, whether these are the DNA molecules of two individuals of the same species or two polyethylene molecules, will not only share "similarities" but will be considered "as equal" in order to identify a species of DNA or a specific poly-ethylene with properties related to an average molecular weight. This general problem of the particular and the general is a very old philosophical question. How can we say that two human beings, although they differ in many charac-teristics, are both members of the same species? What does it mean to cate-gorize a class of objects? Is this categorization by classes always arbitrary? The difficulty of defining a living species and of differentiating it completely from the varieties of members of the same species indicates the philosophical and scien-tific difficulties of such categorizations. These are some of the difficulties that are tackled, from a scientific perspective, by the study of molecular evolution.

This scientific field focuses on how the evolutionary process can be under-stood from DNA, RNA, and proteins. It originated in the 1960s from the con-fluence of the work of biologists with different specialties, from molecular biology to evolutionary and systemic biology, to population genetics, and so on. The study of molecular evolution is based primarily on comparative studies between macromolecules of different species along the evolutionary path. For evolutionary biologists, there is something quite distinctive about the classifi-cation of living organisms through DNA, something that does not apply to any other type of classification. This special "something" follows from the specifi-cally evolutionary idea that there is only one correct genealogical tree for all living organisms, upon which we can base our taxonomical classifications. The

[11] Staudiger, "Die Chemie der hochmolekularem organischen Stoffe im Sinne der Kekuléschen Strukturlehere," as cited in Cerruti, *Bella e Potente. La chimica del Novecento fra scienza e società*, 208.

"substance" that holds this tree together is the DNA molecule, from whose variations all living species have been generated.

There are two ways in which DNA can be related to the birth and transformation of biological species. These two fundamental elements of molecular evolution are point mutations, particularly those that take place in the part of the DNA that encodes for specific proteins, and regulatory mutations. Although a point mutation is a simple substitution of one of the bases of DNA, we see that it can modify the amino acid sequence of a protein if such a modification occurs in the coding part of the DNA. A regulatory mutation, on the other hand, is any type of mutation that occurs in or around a gene and is able to influence the activity or inactivity of that gene itself. The study of point mutations has led to the concept of the "molecular clock" and to the discovery of a type of genetic change known as neutral mutation, that is, a mutation that is neither advantageous nor disadvantageous in itself for the carrier organism.

The molecular clock is used in the study of molecular evolution to estimate the time that has elapsed since the separation between two species, beginning with study of the current existing differences in their DNA profiles. This method is based on the assumption that, if the mutations are random, the DNA transformation occurs with frequencies that are approximately constant over time. If this is true, it is possible to estimate the time that has elapsed since the divergence occurred between two species that descend from the same common ancestor and that, therefore, originally had the same DNA. The molecular clock, however, does not work with equal speed at every point of DNA. The rate of evolution at a DNA site that directly affects the function of a protein is low. A position that is free from functional implications can undergo transformations more frequently. In practice, the speed of evolutionary change at the molecular level is inversely proportional to the number of functional constraints. The greater the number of functional constraints, the slower the speed of molecular evolutionary change.

The revolutionary idea that genetic change is dominated by neutral mutations has helped explain the fact that molecular evolution depends more on the amount of time that passes than on the number of successive generations. If molecular evolution were guided by natural selection, its speed would have to be higher in short-lived species such as insects than in long-lived species such as the superior primates. Base substitutions, on the other hand, accumulate at roughly the same rate in the protein-coding sequences of both evolutionary lines.

Today, moreover, many biologists working with mathematical models of evolutionary processes are proposing that many of the mutations accumulated during molecular evolution do not proceed in a linear way. This is because molecular evolution is characterized by long periods of inactivity interrupted by explosions of change. Research on point mutations has also provided new insights into the phylogenetic relationships between different species. By broadening the investigation to regulatory mutations, we can hope to reach a more detailed understanding of the link between molecular evolution and the evolution of an organism. This is because most of the adaptive evolution at the level of the organism is due to mutations that affect the relative concentration of specific proteins, rather than to mutations that modify the structure of these proteins.[12]

Nowadays, the use of DNA to trace the transformation of living species is sophisticated enough to support, if not replace, historical and paleontological research. For example, the research by Luigi Cavalli-Sforza has identified the waves of migration of modern humans from their origin in Africa to various parts of the world. Another good example would be Svante Pääbo's discoveries concerning extinct hominin genomes and human evolution, for which he was awarded the 2022 Nobel Prize. Work on mitochondrial DNA, which is present only in the egg and not in the fertilizing sperm, has instead allowed us to identify a female generational line taking us all the way back to "mitochondrial Eve," the East African woman who is considered the "mother" of all modern human beings. These findings are just a few of the many examples illustrating the use of DNA to identify and follow the modification of a given living species.

Epigenetics

Although the term *epigenesis* is ancient, it may be related to 18th-century embryological theory, according to which the embryo develops gradually from an undifferentiated seed. However, epigenetics, as the study of epigenesis, was introduced in the 1940s by the English biologist Conrad Hal Waddington. Epigenetics was concerned, on the one hand, with the development of the embryo, which was understood, from this perspective, as a gradual and qualitative change. On the other hand, epigenetics was also concerned with the causes of development, especially those causes involving interacting sets of genes and how such causes led to phenomenal characteristics. The prefix "epi," which means "over" or "upon," was used because these studies went "beyond" the gene. Waddington's approach to epigenetics was systemic and not in line with the simple, linear, cause–effect approach that was then prevalent in molecular biology.

Beginning with the observations that all cells of a multicellular organism have the same genes and that the various cell types that make up an adult organism may differ significantly, some scientists might infer that epigenetics should identify the causes of this process of differentiation. However, epigenetic mechanisms are also identified in single-celled organisms that, although they share the same

---

[12] Wilson, "Le basi molecolari dell'evoluzione."

DNA sequence, can express distinct phenotypes. Therefore, epigenetics involves the study of molecular mechanisms that lead to the establishment of a gene expression program that determines the identity of a specific cell. However, the question that remains is whether epigenesis excludes the programs that are activated or deactivated in response to the presence or absence of an external stimulus. Obviously, the essential point is that a process is considered epigenetic if it is stable and not easily reversible, so that the epigenetic characteristics of a cell are transmitted to the progeny.

Returning to the structure of DNA, its double helix is organized within the cell nucleus, in interaction with specific proteins and in a three-dimensional structure called chromatin. Chromatin is a more or less compact structure, and its degree of compaction regulates the accessibility of RNA polymerase to DNA and can thus modify phenotypic characteristics. We currently know of two general types of modification that alter the conformation of chromatin: (1) Modification of histones (basic proteins that constitute the structural component of chromatin), which constitutes the so-called histone code[13] and (2) DNA modification (such as methylation of specific cytosine residues) that does not alter its base sequence. Study of the effect of these modifications on phenotype is an integral part of epigenetics.

These modifications are different from those in which an influence is exerted by the environment during embryonic life, when the physiology of an individual can be strongly influenced by the intrauterine nutritional conditions. A famous example of this are the people conceived during the great famine of the Dutch Hunger Winter of 1944–1945. In this case, 60 years later, individuals who were prenatally exposed to the famine had less DNA methylation of the imprinted *IGF2* gene compared to their unexposed, same-sex siblings. Thus, the individuals who had been exposed to famine while in the embryonic developmental stage displayed a 10% increase in mortality after 68 years.

An increasing number of experimental evidence suggests the possibility of transmitting epimutations that are caused by the environment to one's offspring. In such a case, these epimutations would constitute a transmission of acquired characteristics, which suggests that some form of Lamarckian transmission is also possible in complex animals, such as mammals. In its broadest sense, epigenetics rewrites the genetic paradigm because it postulates that intergenerational information cannot be referred only to programmed gene expression. Rather, the epigenetic paradigm seeks to demonstrate that embryonic development, at least, can introduce innovations that are transmissible to the offspring. In this case, the epigenetic vision transcends the simple study of those heritable changes in gene function that occur in the absence of mutational variations in

---

[13] Jenuwein and Allis, "Translating the Histone Code."

DNA. Even the meaning of environment is altered when we move from a genetic to an epigenetic perspective. In epigenetics, the environment goes from being a simple selector of a predetermined program to becoming a co-producer of such a program.

Following Jan Baedke,[14] we assert that epigenetics has two different dimensions: the molecular dimension and the ecological and evolutionary dimension. Many biologists and philosophers of biology believe that an epigenetics that is closely linked to the molecular approach is reductionist, while one that is linked to an ecological and evolutionary approach is systemic. However, we believe that both epigenetic approaches, as well as classic genetic approaches, must move in the systemic direction proposed in this book. The molecular approach to biological questions does not intend to reduce biology to chemistry, just as the atomic approach to chemistry cannot reduce chemistry to physics. This general point was made in the introduction to this book and does not require further reiteration here. The molecular approach to epigenetics also poses minor problems for the philosophical approach of classical genetics, since it relies on the same molecular level of explanation as DNA. On the other hand, both in genetics and epigenetics, the ecological and evolutionary approach, framed in systemic rather than reductionist terms, must always be linked to the molecular explanations of processes.

This being said, the so-called Lamarckism of epigenetics deserves a separate philosophical treatment. For decades, biology has considered the relationship between the environment and the genetic code to be exclusively of a Darwinian type, proceeding from a random mutation of the molecular part (DNA) to its indirect selection through the phenotype, in order to maximize the adaptation of the organism to its environment. We are not yet in a position to say whether this relationship should be made more complex by some form of Lamarckian selection. However, we can affirm that the environment and the system are both complex systems and are in systemic and complex interaction with each other. They can, therefore, alter each other. Chapter 11 already discussed Morin's view that, in addition to the competitive aspect of natural selection, we must add a complementary component in evolution in which the selection process enters a recursive ring with that of integration. The work of the American biologist Lynn Margulis comes to mind here, but a discussion of this work would go beyond the scope of this chapter, although some of her research is discussed in the next section, in addition to some aspects of Morin's evolutionary theory.

Before moving on to the next section, however, we reiterate that, according to the chemical approach to biology that we adopt in this book, we believe that there is no clear philosophical distinction between classical genetics and

---

[14] Baedke, *Above the Gene, Beyond Biology: Toward a Philosophy of Epigenetics.*

epigenetics. Both fields of study describe complex systems in complex interaction with each other, and both fields must adopt molecular and ecological systemic perspectives. However, the different role assumed by the environment in the epigenetic paradigm retains broad epistemological interest.

Metagenomics

The human body consists of about 10^{13} cells, the microorganisms that reside in the human body are estimated at around 10^{14}, and the major bacteria groups in the human gut are about 1000. Therefore, the microbiome, the genome of microbial communities living in the human body, contains more genes than those of the human genome. Metagenomics studies the microbiome collectively, without separating it into the different species of microorganisms.[15] Lynn Margulis's generally accepted endosymbiont hypothesis[16] postulates that complex cells (eukaryotic cells) were generated from the fusion of primitive organisms and that the mitochondria and plasmids that are now part of these cells were originally autonomous organisms. Therefore, the role of symbiosis in the fusion of organisms is considered as established fact.

In the case of the human body, many processes that occur within it depend on the microbes that colonize it. The vast majority of these microbes are not only inoffensive to humans but actually provide some essential functions in the body, such as the digestion of food and the counteracting of toxins. Therefore, they act as a support in fighting against diseases that are caused by other microbes that can invade the body. Nobel laureate Joshua Lederberg coined the term *super organism* to describe the set of human cells along with those microbes that occupy the human body.

In addition to such microorganisms, there is also a huge number of viruses in the body that are not usually recognized as independent life forms but that use human "instrumentation" to reproduce. These bacteria can communicate with each other using the chemical language of production and perception of specific signal molecules. They have evolved systems that allow them to "count" and express certain genes only when their population reaches a certain size or "quorum" (quorum sensing). This phenomenon allows a microbial population to regulate, in a coordinated way, the expression of a series of genes according to the density of its population. The microbial "genetic patrimony" can be considered part of the genetic patrimony that is available to human beings because it carries

[15] Thurnbaugh, Ley, Hamady, Fraser-Liggett, Knight, and Gordon, "The Human Microbiome Project Exploring the Microbial Part of Ourselves in a Changing World."
[16] Sagal, " On the Origin of Mitosing Cells."

out essential active functions for us. Once again, the concept of the genetic patrimony of a species "widens," making its definition and function increasingly complex and systemic.

We conclude this chapter with a citation that is dedicated to the memory of
Lynn Margulis, who was the master architect for rethinking biology in terms of
interacting consortia:

> The notion of the "biological individual" is crucial to studies of genetics, im-
> munology, evolution, development, anatomy, and physiology. Each of these
> biological subdisciplines has a specific conception of individuality, which has
> historically provided conceptual contexts for integrating newly acquired data.
> During the past decade, nucleic acid analysis, especially genomic sequencing
> and high-throughput RNA techniques, has challenged each of these disci-
> plinary definitions by finding significant interactions of animals and plants
> with symbiotic microorganisms that disrupt the boundaries that heretofore
> had characterized the biological individual. Animals cannot be considered
> individuals by anatomical or physiological criteria because a diversity of
> symbionts are both present and functional in completing metabolic pathways
> and serving other physiological functions. Similarly, these new studies have
> shown that animal development is incomplete without symbionts. Symbionts
> also constitute a second mode of genetic inheritance, providing selectable
> genetic variation for natural selection. The immune system also develops,
> in part, in dialogue with symbionts and thereby functions as a mechanism
> for integrating microbes into the animal-cell community. Recognizing the
> "holobiont"—the multicellular eukaryote plus its colonies of persistent
> symbionts—as a critically important unit of anatomy, development, physi-
> ology, immunology, and evolution opens up new investigative avenues and
> conceptually challenges the ways in which the biological subdisciplines have
> heretofore characterized living entities.[17]

---

[17] Gilbert, Sapp, and Tauber, "A Symbiotic View of Life: We Have Never Been Individuals."

13

Cellular Chemism

Although not widely used in the literature, the term *chemism* is not new. In general, it can be linked to an explanation that refers to the chemical entities and processes of a material body or a phenomenon. The 19th-century philosopher Georg Wilhelm Friedrich Hegel used the term *chemism* in opposition to *mechanism*. For Hegel, in mechanism, the relationship between two objects is a relative determination of space–time. In chemism, on the other hand, two substances are in a more intimate relationship through their "nature." Hegel noted that chemism is often included under mechanism and both are contrasted with teleology, that is, with the purpose of things. However, although Hegel agreed that both mechanism and chemism exclude teleology, he sought to distinguish between them. In Part I of the *Encyclopedia of the Philosophical Sciences*, he states that

> [t]he object, in the form of mechanism, is primarily only an indifferent reference to self, while the chemical object is seen to be completely in reference to something else. No doubt even in mechanism, as it develops itself, there spring up references to something else: but the nexus of mechanical objects with one another is at first only an external nexus, so that the objects in connection with one another still retain the semblance of independence. . . . The case is quite different in chemism. Objects chemically biased are what they are expressly by that bias alone. Hence they are the absolute impulse towards integration by and in one another.[1]

As Johann W. Goethe's book *Elective Affinities* demonstrates, the chemical concept of affinity had acquired merit at the beginning of the 19th century, and this seems to be what Hegel means by "the absolute impulse towards integration by and in one another."

In biochemistry today, the term *chemism* designates the series of phenomena determined by chemical action, and it is in this sense that we use it in this chapter in which we deal with intracellular (chemical) processes. Although the term most commonly used to indicate the set of processes (i.e., chemical reactions) that take place within cells is *metabolism*, we prefer to use the term *chemism* in

---

[1] Hegel, *Encyclopedia of the Philosophical Sciences, Part I: The Logic*, 274.

From the Atom to Living Systems. Marina Paola Banchetti-Robino and Giovanni Villani, Oxford University Press.
© Oxford University Press 2023. DOI: 10.1093/oso/9780197598900.003.0014

this chapter in order to highlight the chemical nature of our description of what occurs in the cell.

In this book generally, but more specifically in the last few chapters, we have shown why the term *chemism* should be valued. This term, in fact, immediately and clearly highlights the principal, though not unique, approach according to which the living being must be "explained," that is, the chemical approach. The matter forming living organisms and the processes taking place within them are "understood" through chemical language. We have stressed that this approach to reading the material world is principal, but not unique, and there are, in fact, different perspectives from which one can look at both inanimate and conscious living organisms. There is, for example, the physical approach of cellular processes or, at least, of parts of these processes. This approach does not make reference to chemical entities, such as molecules and macromolecules, but to potentials, to electric and magnetic fields, and to other physical concepts.

There is another perspective for describing what happens within a living organism, which is often referred to as the information approach. This latter perspective is sometimes used to generalize the chemical approach and render possible a concept of life as independent of its material expression. Such a generalization is not in opposition to the chemical perspective of life and could be useful if a different (and not chemically interpretable) form of life were to be found on other celestial bodies.

We have also applied both the physical and the information perspective, and we will continue to apply them in specific cases, such as when we say that DNA represents a "code" that carries information between generations of a living species. In this case, the chemical nature of this macromolecule takes second place and foregrounds the genetic information, its constancy, its variability, and so on. Even within this perspective, however, the chemical nature of DNA and the way in which this molecule can store and transfer information will remain basic concerns.

The Cell: The Fundamental Unit of Living Beings

In this section, we will focus on cellular *chemism* by assuming that the cell is the fundamental unit of life. Thus, in dealing with the cell, we deal with living beings in general. However, the identification of the cell with the fundamental unity of the living requires a philosophical clarification and some scientific considerations, beginning with the determination of what a living being is.

The philosophical clarification we are offering concerns the general aspects of the relationship between the cell and the multicellular organism that encompasses it. We say philosophical rather than scientific here because, when examining these aspects, we must also consider the mereological relationship between the parts and the whole in a system. The systemic perspective gives relative meaning to whether the emphasis is placed on the multicellular system (pluricellular organism) or on the single cell. Such an emphasis generally depends on many factors, particularly on the motivation of the study. The same subject matter can, in fact, be approached and addressed from different levels of complexity.

Consider the example of a disease, which is a global dysfunction of the organism and as such is closely related to the entire organism, in both its material and psychological aspects. Good examples of this are psychosomatic diseases but also more specific problems such as those of the respiratory system. In such cases, the emphasis is placed on a set of connected organs that form such a system. Considering the disease as respiratory, however, does not exclude its psychosomatic aspects, as is well documented for many asthmatic conditions. Asthma, however, can still be described and linked to a malfunction of the lungs.

The specification process can progress further and connect the disease to lung tissue, lung cells, or even to genetic information that activates or disables a particular protein that is involved in the breathing process. All these levels are proper and important for understanding the disease, and none of them is completely reducible to a "deeper" level (in the philosophical but also scientific, sense). In practice, the question of what a disease is, or even how a disease can be cured, can be answered in multiple ways that are all appropriate and correct, once one has specified the level of analysis at which one is working.

Returning to the cell and its function in a multicellular organism, we point out that historically there have been two extreme and contrasting perspectives on this issue: biological atomism, which was completely reductionistic, and the organismal theory of life, which was completely holistic. Biological atomism assumed that every living being was composed of elementary units to which all the properties that characterized the "living" could be attributed. Therefore, the activities of the body could be conceived as resulting from the activities of these units. The clearest example of this approach considers the cell as the fundamental unit of life and the pluricellular organism as an aggregation of autonomous living units.

The organismal theory, on the other hand, assumes that the entire organism, rather than its cells, is the primary unit of life. Thus, the organism is considered to the be the cause of cellular constitution, not its product. Rudolf Virchow,[2] one of the originators of the concept of cell, agrees both with biologists who purport to analyze life up to the last molecular movements and with biologists who attribute life to the entire organism. These two opposing orientations can be

---

[2] Virchow, *Vecchio e nuovo vitalismo*, 99–158.

unified by considering the cell as a fundamental datum of biological theory and as a vital entity, taking into account that an organism is composed of cells and that it is only in the cell that molecules can be composed into a true living unit.

The holistic organismal standpoint, later known as the organismal theory, was developed at the end of the 19th and the beginning of the 20th centuries. Its critique of biological atomism was based on the recognition that, since multicellular organisms are individuals rather than communities of individuals, their evolutionary origin has to be located not in the association of many cells but in a single organism with specialized parts with different functions. For William Emerson Ritter, the primary unit of life and, thus, the true "biological atom" is not the cell but the entire organism.[3] The organism is considered the cause rather than the effect of its cellular constitution. Of course, cells are considered important. However, one cannot make sense of cellular structure, function, and organization from outside the perspective of the entire organism.

In practice, organismal theory assumed that the organism was made of cells but that its biology was not reducible to cell biology or to the biology of its constituents. Our systemic perspective agrees with this assumption, and, like him, we provide the following analogy to chemical processes to support this view. Just as oxygen and hydrogen atoms lose their independent identities when they combine to form a water molecule, living cells fuse their separate individualities when they collectively constitute an organism. One of the main exponents of this organismic critique of cellular atomism was the English biologist Clifford Dobell, who also refused to accept the designation of unicellular to describe protists, organisms that we consider to be formed by a single cell. He said that they should be considered as "a-cellular" organisms.[4] Although this a-cellular designation has not been adopted in biology, it expresses very clearly the difference between a cell and a pluricellular organism.

Regarding scientific considerations, the meaning of "living" and of "life" has changed in various ways over the centuries, and summarizing all these meanings is certainly beyond the scope of this book. We refer the interested reader to André Pichot's excellent *Histoire de la notion de vie*.[5] For our purposes here, we simply note that one method of defining the living is to list the essential properties that characterize living things, leaving aside the possible uncertainty regarding whether or not viruses should be included. In *Chance and Necessity*, Jacques Monod conducts a thought experiment that allows him to identify three of these essential characteristics. In recounting the conclusions that he reached,

[3] Ritter, *The Unity of the Organism: Or the Organismal Conception of Life*.
[4] Dobell, "The Principles of Protistology," as cited in John O. Corliss, "The Protozoon and the Cell: A Brief Twentieth Century Overview."
[5] Pichot, *Histoire de la notion de vie*.

Monod uses the term *teleonomy*, which he invented to mean "objects with a purpose or a project."[6]:

This imaginary experiment has had no other aim than to compel us to "rediscover" the more general properties which characterize living beings and distinguish them from the rest of the universe. Let us now admit to a familiarity with modern biology, so as to go on to analyze more closely and to try to define more precisely, if possible quantitatively, the properties in question. We have found three: teleonomy, autonomous morphogenesis, and reproductive invariance.[7]

In a more current, though less influential, book on the same problem of defining life, Paul Agutter and Denys Wheatley address cellular life in the following way:

We suggest that "livingness" is characterized at the cellular level by a three-way dialogue among:
- the internal state;
- the set of all responses to external stimuli; and
- the pattern of gene expression.

"Internal state" encompasses cell structure, metabolism and internal transport, locked together by reciprocal dependence at any moment. "Responses to stimuli" include all cell's signaling pathways. "Pattern of gene expression" means the rate (zero or finite) at which *each* gene is being transcribed. . . . Fluctuations in the internal state affect stimulus-response and gene expression more or less immediately (t1). Altered responses to stimuli affect internal state and gene expression after a slight delay (t2). Changes in the gene expression pattern affect internal state and stimulus response after a longer delay (t3).[8]

Regardless of whether or not we agree with Agutter and Wheatley on the general properties that define life, we are interested in the notion that the concept of life always requires a cyclical relationship or a three-way dialogue of its defining properties. In this specific case, after choosing the three sets of fundamental properties, the "internal state," the relational "responses to external stimuli," and the "pattern of gene expression" (the need to have a "program and the way it is decoded"), Agutter and Wheatley immediately rush to the conclusion that life is possible if and only if these properties are present all together and in close interrelation. From our systemic point of view, however, what must be highlighted is the interrelation of the properties rather than the list of individual properties that must be present. According to this definition of life, the cell is a fundamental

---

[6] Monod, *Chance and Necessity. An Essay on the Natural Philosophy of Modern Biology*, 13.
[7] Ibid.
[8] Agutter and Wheatley, *About Life: Concept in Modern Biology*, 91.

unit of life in the same way as a complex organism, while viruses do not fall into the category of living beings.

We have already discussed the philosophical and scientific importance of these cycles of relations in the introduction, and we will address them later as well. In what follows, we will consider the three sets of properties discussed above and show their relationships in detail, in order better to analyze and highlight the concept of cellular *chemism* that is the central focus of this chapter.

The Internal State of the Cell

For an entity to exist, be it living or nonliving, there must be a surface (in the broad sense) that separates it from other entities. In the case of the cell, this surface is the cell membrane, which is materially well defined. The membrane encloses the cell, separating it from the environment, but it is also permeable to some substances, thus allowing for the exchange of matter; to some external stimuli, thus allowing the exchange of information; and to the flow of energy, thus allowing for the exchange of energy. In thermodynamics, this type of system is defined as open and is opposed to the isolated system in which there is no ability to exchange either matter or energy with the environment outside the system. An open system is also opposed to a closed system in which there can be exchange energy but not exchange of matter with the environment outside the system. However, thermodynamics is not concerned with the information that can enter or leave a system and that is attributable to exchanges of matter and/or energy of that system with its environment.

Once we have enclosed the portion of the world we call the system, we characterize it with its dynamic state, understood in the broad sense, which determines the specific conditions of that part of reality. Following Agutter and Wheatley, we can say that

"Internal state" denotes the quantities of all the cell's ingredients at a particular moment, their organization in space, the sum total of the metabolic events taking place, the directions of all the transport processes and what they are transporting, and the ways in which these different features interlock. "Internal" indicates the overall situation within *one cell*, not what might be taking place outside it or in other cells. 'State' is fairly non-committal, but connotes a definable, reasonably stable situation.[9]

A clarification is appropriate here. We believe that the concept of internal state of a living system should be extended to the "set of internal states" of the open system that allow the persistence of the property called life. This means,

---

[9] Ibid., 53.

for example, that although the normal state of the human body is a temperature of about 36.5 ° C, having a temperature of 35 or 39 °C does not render a person "less human." The presence of a set of similar states in which the set of human properties varies little helps us to understand what has changed in a living being when it has died. The answer is that the open system is no longer the set of states associated with the global property of life and is now in a state that leads to the closure of the system and the disappearance of this global property. This means that the open system stops functioning as such and closes itself. Connecting life to a set of states of an open system with a specific global property allows us both to overcome the notion of life as a single emergent property and to follow the transformation of the system, as well as the loss of this property. This topic is mentioned only briefly here, but it certainly deserves to be deepened further.

An essential point of the cellular state is that, although it is dynamic and can thus be modified over time, it represents an optimal situation of the cell that persists in its principal characteristics. As previously discussed, in the second part of the 19th century, Claude Bernard introduced the concept of *milieu intérieur*, which is fundamental in modern biology. The constancy of the *milieu intérieur* is the condition for liberating a living organism from the external environment in which it lives. In contemporary biology, this concept is expressed by the term *homeostasis*, that is, the system's ability to keep constant an important series of physical parameters that must be regulated with extreme care in order for the cell to remain alive.

The term *homeostasis* was introduced in 1929 by Walter Cannon, who also referred to the strategies by which animals are able to maintain this constancy. The simplest strategy involves the storage of surplus stocks in times of abundance, either by simple accumulation in selected tissues (e.g., water in the muscles or skin) or by conversion to a different form (e.g., glucose to glycogen). A second type of homeostasis involves altering the speed of continuous processes (e.g., changing the speed of blood flow by changing the size of the capillaries to maintain uniform temperature). Finally, Cannon noted that such control mechanisms could be regulated by the autonomic nervous system.

If we inquire about the processes that occur within the cell, we note that, even if the incessant process of transport within the cell is neglected, two types of processes continuously occur: (1) cell structures are mounted and dismantled and (2) many chains (pathways) or cycles of chemical reactions are always in progress. This brings us to an examination of the components of the internal state of a cell, which we divide into four groups: (1) the substances present, (2) the organization of the individual cellular parts, (3) the chemical reactions that take place in the cell, and (4) the exchanges (in the broad sense) between the cell and its external environment.

Let us address in more detail the first two groups of components of the internal state of a cell: the substances present and the organization of cellular parts.

(1) The main material constituent of the cell is water, a chemical substance classified as inorganic. Water is the most abundant substance in the cell, making up about 70% of its weight. It is also the medium within which chemical reactions occur and is important for thermal equilibrium, contributing to the constant maintenance of cell temperature. Other inorganic chemicals in the cell are certain mineral salts that present as electrolytes and are dissolved in water in their ionic form. The organic substances present in the cell, however, are divided into two large categories: small organic molecules (conventionally those with less than 50 atoms) and macromolecules. The first group is primarily made up of simple sugars (carbohydrates), fats (lipids), amino acids (basic units of proteins), nucleotides (basic units of nucleic acids), and vitamins (essential substances that are present only in small quantities). The second group includes proteins (such as enzymes), more complex carbohydrates (such as glycogen), and nucleic acids (RNA, DNA).

(2) With regard to cell organization, a first subdivision is in prokaryotic and eukaryotic cells, depending on whether they have or do not have a well-characterized cell nucleus, although sometimes they have more than one. The cell, delimited by the cellular membrane, contains within it the cytoplasm in which numerous cellular organelles are present: the endoplasmic reticulum, the ribosomes, the Golgi apparatus, the mitochondria, the lysosomes, the centrioles, and, for vegetable cells, the plastids. Cells use a variety of organelles to perform their main functions.

We will not discuss the individual specialized cell pieces but will only focus on the membranes that identify and separate them because today's research increasingly shows that important cellular activities involve events that occur on the surfaces of a complex system of membranes. In addition to surrounding the surface of the cell and its nucleus, the membranes determine the fundamental compartmentalization of the cytoplasm and divide the cell into individual organelles, segregating within them specific components and processes. We can certainly say that the cell membrane played an essential role in the evolution of cells, from the first cell that separated a small amount of aqueous solution from the rest of the environment to the cells that exist today. For this reason, and by way of example, we describe in some detail the cell membrane, even though we understand that each membrane in the cell has typical characteristics that specify its function, in addition to common characteristics.

The cell membrane consists of lipids, proteins, and, to a small extent, carbohydrates usually directed toward the outer surface of the cell. The amount of the components is, however, linked to the membrane of specific cells. For example, the membrane of neurons, which are the fundamental cells of the nervous system, is covered by a myelin sheath that is formed mainly by a lipid component, which plays the role of electrical insulator. In general, the cell membrane is formed by a double lipid layer, interrupted by proteins that can protrude either from both sides of the cell or from one side only (either toward the inside of the cell or toward the outside).

The double lipid layer, however, is not a static structure, since the lipid molecules can move from one area to another and also pass from one layer to another (*fluid mosaic model*). The first thing to highlight is that the cell membrane is not a passive barrier but an active selection system. The double lipid layer of the membrane allows free passage only to water, to O_2, to gaseous CO_2, and to small fat-soluble molecules such as ammonia, urea, and ethanol. There are specific membrane proteins that ensure the active transport of water-soluble ions and molecules through the membrane. As already established, the cell membrane contains proteins that are fundamental for determining the transport functions of substances in two directions (input and output), for the passage of the signal and, in the case of substances that adhere to the surface, for the function of recognition.

From a general and philosophical point of view, the cell membrane is what concomitantly separates and unites the cellular system. This seems like a contradiction because the two functions of separating and uniting are generally considered to be in opposition, and it is generally believed that something either unites or separates. However, from a systemic and cyclical perspective, these two functions are not in opposition to each other but are two sides of the same coin.

Let us now address in more detail the third group of components of the internal state of a cell: the chemical reactions that take place in a cell.

(3) Thousands of chemical reactions occur in the cell, and all have some form of regulation. The need to regulate every aspect of cellular metabolism is linked to the complexity of the sequence of reactions. We distinguish metabolic pathways with precise tasks in the cell economy but all these pathways are interconnected, forming a complex network. The cell is able to pass from one metabolic path to another when the external environment changes, for example from anaerobic conditions (without oxygen) to aerobic conditions (in the presence of oxygen). In practice, cells have the metabolic capacity to activate preferentially one route over another in order to generate the products needed for cell life within the changing conditions of the external environment.

Some metabolic pathways are linear, and some are branched; that is, they start from a chemical and produce one or more distinct substances, at the end of a certain number of related steps. The main metabolic pathways, however, are cyclical. Some metabolic pathways serve to reduce complex molecules into simpler molecules and to release energy. All processes of this type are grouped as catabolic. Some metabolic pathways use simpler molecules and energy to build more complex molecules (anabolic processes). The anabolic and catabolic aspects of metabolism are closely and harmoniously interconnected.

This does not mean that there is a biunivocal (one-to-one) relationship between these two aspects of metabolism. Their independence is remarkable with regard to the metabolic pathways followed, to the regulatory processes, and to the intracellular location of such processes. An example is the degradation of fatty acids in liver cells that takes place in mitochondria, while their synthesis takes place in the cytosol, which is the liquid and unstructured component of the cytoplasm. However, this independence has exceptions such as, for example, in the Krebs cycle where the demolition process is inextricably linked to the typically anabolic process of energy production. The Krebs cycle, moreover, is an essential hub of all cell metabolism, and its activity is closely coordinated with those of the other metabolic pathways, something that we will discuss more later in this chapter. Once again, a closed succession of interrelated steps or cycles (in this case, one of reactions) is essential to life, and this is a central field of study for systemics.

We shall now address in more detail the fourth group of components of the internal state of a cell: the exchanges that occur between the cell and its external environment.

(4) We have established that the cell is an open system, and, thus, the exchange of material from and to the outside is fundamental. Every living cell must obtain the material it uses in the internal biosynthesis from the surrounding environment, and it must release the waste of the metabolism into the surrounding environment. Some solutes can pass according to the gradient of concentration, that is, in the direction that goes from the more concentrated environment to the less concentrated one. We could say, in a metaphorical sense, that they pass in the natural thermodynamic direction, while others are pumped against the concentration gradient and thus require a waste of energy. Ions (charged species) can move through the membrane in so-called ion channels, such as that for potassium or sodium, and such channels may be sensitive to physical properties such as temperature or mechanical stresses.

222 FROM THE ATOM TO LIVING SYSTEMS

For example, the potassium ion (K^+) channel consists of four identical units that pass through the membrane forming a cone-shaped structure. This channel allows the passage of a single row of nonhydrated ions, which have lost the water molecules surrounding them in solution, and such dehydration takes place inside the channel. On the inner and outer surface of the cell membrane, there are negatively charged particles that increase around the concentration of positive ions. The discrimination between the K^+ and the Na^+ ion occurs due to internal channel interactions that are established for a larger ion, such as K^+, but not for the Na^+ ion. Therefore, the ability to discriminate ions depends upon the charge of the external parts of the channel, the size of the channel, and the interactions inside the channel.

This analysis of ion channels leads us to the following consideration. We often tend to think of the interactions between chemical species as something geometric. For example, in the lock-and-key model of enzymatic activity that we discuss later in this chapter, we employ expressions such as the right orientation between the species in interactions. However, chemical entities are not geometric in nature. They are not large or small balls, nor are they triangles or squares. The chemical activity of these entities depends on the spatial distribution of internal electric charges and is different from part to part within the system. The molecule and the macromolecule can, therefore, be characterized in a qualitatively different way in different parts. In visual terms, this would be analogous to saying that every molecule and macromolecule not only has a different shape but also different colors for different parts.

The Cell's Ability to Know and React to Its Environment

From the systemic perspective, the ability of a system to interact with its environment as a global unit is one of its fundamental features. In the introduction, we discussed the importance of the concept of an open system, a type of system that is easily linked to living systems but that is not exclusive to them. The ability to "know" the extracellular environment and be able to react to it is one of the essential characteristics of life. This is true both for cells of pluricellular organisms and for unicellular organisms. The bacterial cells "read" acidity and osmotic pressure and "feel" the presence of useful/harmful substances, such as oxygen or nutrients, and the presence of predators and competitors. All of these stimuli induce specific responses, such as distancing or approaching the source of the signal. In pluricellular organisms, besides the above stimuli, there are also stimuli that the cells of the same organism exchange with each other.

Hormones are called chemical messengers that transmit information from one cell or group of cells to another cell or group of cells. Hormones have the task of modulating the metabolism and/or the activity of the cells of the same organism and are one of the elements that constitute the organism as a whole, since they "connect" and "inform" the single parts. In all these cases, the chemical signal represents the information that the cell receives through specific receptors, which can be analyzed in chemical terms, and the response of the cell always implies a chemical process. The technical term for this process is transduction of the signal. This is a universal process in all living beings, and it is "readable" in chemical terms.

For each type of organism, the signal transduction pathways are specific and sensitive, and, in pluricellular organisms, they are different for different cell types. The specificity of transduction comes from the "molecular recognition" between the molecules of the signal and those of the receptors. The sensitivity of the transduction pathways depends on various factors that are both specific to the molecules directly involved, either by the presence or absence of other molecules in the environment or by the amplification of the signal by enzymatic reactions. The sensitivity to the signal is also changeable over time. For example, there is a desensitization of the receptor if the signal continues for too long.

Finally, cells have a capacity for integration, that is, the ability to add more signals received in order to give an appropriate global response. Although there are thousands of signal transduction routes, the molecular mechanisms that govern the process include only a dozen basic substances. As a result, signal transduction processes are often identified by the name of one or more receptor molecules (e.g., receptors with tyrosine kinase activity), just to highlight the molecular performer of such a process. Therefore, even in their diversity and specificity signal transduction processes can be schematized globally according to their common characteristics: (1) all cells have specific and sensitive transduction mechanisms; (2) signals interact via molecular cell receptors; (3) the activated receptor interacts with cellular chemism by producing a second signal or varying the activity of a protein; (4) the metabolic activity of the cell changes and; (5) the transduction process terminates by returning the cell to the initial condition.

From a philosophical point of view, the characteristic of "knowing" your environment, of "selecting" the positive and negative aspects, and of knowing how to "respond" in different ways to the external perturbations also configure the simplest cell as something extremely complex, with characteristics that were once thought to be present only in the so-called superior animals, if not only in human beings. To be clear however, although we have described these fundamental processes using linguistic terms, such as *read, know,* and *select answers,* these extraordinarily complex processes are fundamentally based upon "simple" chemical interactions and reactions.

A Code in the Cell and the Processes Connected to It

In this chapter, we will not examine the details of the genetic code or its material basis in the DNA macromolecule because this special macromolecule was already extensively discussed in Chapter 12. As has been discussed, DNA performs two essential roles for life: it participates actively and as a protagonist in cellular chemism, and it allows the propagation of information essential to life and its specificities beyond the individual life of the organism. Both of these functions were examined in detail in the previous chapter, along with the historical development of our understanding of this macromolecule. In this chapter, therefore, we will only highlight the informational aspect of DNA and why the discussion of the genetic code is so significant.

When we speak of the *genetic code*, we mean by this the set of rules that govern how the information that is present in a specific sequence of DNA is translated for the synthesis of a protein in the cell. Thus, the process that leads from DNA to protein has two aspects: the reserve of information (DNA) and its translations or clarification. We will now quote at length from Dawkins's *The Blind Watchmaker*, in which he explains the relationship between these two aspects by using the analogy of a cake recipe:

> A recipe in a cookery book is not, in any sense, a blueprint for the cake that will finally emerge from the oven. . . . a recipe is not a scale model, not a description of a finished cake, not in any sense a point-for-point representation. It is a set of *instructions* which, if obeyed in the right order, will result in a cake. A true one-dimensionally coded blueprint of a cake would consist of a series of scans through the cake, as though a skewer were passed through it repeatedly in an orderly sequence across and down the cake. At millimeter intervals the immediate surroundings of the skewer's point would be recorded in code; for instance, the exact coordinates of every currant and crumb would be retrievable from the serial data. There would be strict one-to-one mapping between each bit of the cake and a corresponding bit of the blueprint. Obviously this is nothing like a real recipe. . . . Now, we don't yet understand everything, or even most things, about how animals develop from fertilized eggs. Nevertheless, the indications are very strong that the genes are much more like a recipe than like a blueprint. Indeed, the recipe analogy is really rather a good one, while the blueprint analogy, although it is often unthinkingly used in elementary textbooks, especially recent ones, is wrong in almost every particular. Embryonic development is a process. It is an orderly sequence of events, like the procedure for making a cake, except that there are millions more steps in the process and different steps are going on simultaneously in many different parts of the "dish". Most of the steps involve cell multiplication, generating prodigious numbers

of cells, some of which die, others of which join up with each other to form organs, tissues and other many-celled structures. . . . Precisely which genes are switched on in any one cell at any one time depends on chemical conditions in that cell. This, in turn, depends upon past conditions in that part of the embryo. Moreover, the effect that a gene has when it *is* turned on depends upon what there is, in the local part of the embryo, to have an effect on. A gene turned on in cells at the base of the spinal cord in the third week of development will have a totally different effect from the same gene turned on in cells of the shoulder in the sixteenth week of development. So, the effect, if any that a gene has is *not* a simple property of the gene itself, but is a property of the gene in interaction with the recent history of its local surroundings in the embryo. This makes nonsense of the idea that the genes are anything like a blueprint for a body. . . . The genes, taken together, can be seen as a set of instructions for carrying out a process, just as the words of a recipe, taken together, are a set of instructions for carrying out a process. . . . Perhaps the best way to see this is to go back to the recipe analogy. . . . But now suppose we change one word in the recipe, for instance, suppose "baking-powder" is deleted or is changed to "yeast." We bake 100 cakes according to the new version of the recipe, and 100 cakes according to the old version of the recipe. There is a key difference between the two sets of 100 cakes, and this *difference is* due to a one word difference in the recipes. . . . There will be a reliable, identifiable difference between cakes baked according to the original version and the "mutated" versions of the recipe, even though there is no particular "bit" of any cake that corresponds to the words in question. This is a good analogy for what happens when a gene mutates.[10]

Here, Dawkins begins by differentiating a point mapping of a cake from the recipe for that cake. These are two very different things. In the point mapping, we have a scale reproduction of the object. In the recipe, we have the description of a procedure to obtain the cake. By analogy, Dawkins claims that DNA is much more like a recipe than a point mapping. Embryonic development governed by DNA is a process, an ordered sequence of events, much like the procedure for making a cake, with the difference being that embryonic development involves millions of steps while making a cake involves only a dozen steps at most.

One of the essential points made by Dawkins is that, in processes involving DNA, not all genes are active in a cell at a given time. In fact, which genes are active depends upon the chemical conditions present in that cell at that particular time and also upon the cell's previous history. Dawkins also extends the analogy effectively by comparing gene mutation in DNA with what will occur if the cake

[10] Dawkins, *The Blind Watchmaker: Why the Evidence of Evolution Reveals a Universe without Design*, 295–297.

recipe is changed at any point. From this instructive analogy, we can better understand what is meant by genetic code, as well as the relationship of that code with the information that is being transported.

Two Classes of Chemical Reactions in Cells

There are many types of molecules and macromolecules in a cell, and there are an estimated 100,000 different chemical reactions that transform them. In addition to regulating the substances that enter and leave a cell, in order for the cellular environment to remain constant, it is necessary to spatially and temporally regulate the production and degradation of the substances present at any moment in the cell. Moreover, since the different reactions can interfere with each other, the specific reactions must take place in separate space zones and with the correct timing. All these factors converge in the chemical organization of the cell.

Since the cellular reaction environment · is different for molecules and macromolecules, we shall begin by focusing on the dimensions involved in each case. First, let us examine the size of a "classic" cell. The simplest type of cell (prokaryotic) is about the size of a micron (10^{-6} m), while a eukaryotic cell is about ten times larger (10^{-5} m). Since the cell is not a simple solid with similar size in the three spatial dimensions, we must instead examine the area enclosed by the cell membrane.

A prokaryotic cell has an area of 10^{-12} m^2, while a eukaryotic cell has an area of 10^{-10} m^2. However, the size of the molecules is in the order of some Ångström (10^{-10} m), and, thus, these have an area of about 10^{-19} m^2. Therefore, the ratio between the cellular and molecular areas is 10^7 and 10^5 for prokaryotic and eukaryotic cells, respectively. A generic macromolecule is of the size of 10^{-8} m, with an area of 10^{-16} m^2. Even if we neglect their aggregations, we see that the ratio between the cell area and the macromolecular area is not only much smaller than the same cell/molecule ratio, but it is 10^4 and 10^2 for prokaryotic and eukaryotic cells, respectively.

These relationships of areas show that the molecules can easily migrate inside a cell because they find much free space, while, for the macromolecules, the cellular environment is "crowded." For molecular reactions, we will need to employ normal statistical laws. On the other hand, for macromolecular reactions, we must consider their individuality and what occurs on the single macromolecule. As Elizabeth Kutter states: "It is clear that any real understanding of what is happening in living cells must take the complex interactions between macromolecules into consideration, along with individual enzyme properties and the effects of intracellular concentrations of ions and other small ligands."[11]

---

[11] Kutter, "Analysis of Bacteriophage T4 Based on the Completed DNA Sequence," 27.

As already mentioned, the metabolic pathways and/or reaction cycles are an important aspect of molecular chemical reactions. In practice, the whole group of reactions is important, rather than simply the individual reactions. To illustrate this point, let us examine the citric acid reaction cycle known as the Krebs cycle, named after the 20th-century German biologist, Hans Krebs. This will help to clarify the way in which many molecular chemical reactions are coordinated and how some compounds are at the crossroads of several metabolic pathways.

The Krebs cycle is part of the broader process of cellular respiration, the set of chemical processes that consumes oxygen and produces carbon dioxide. Most eukaryotic cells and many bacteria that live in the presence of oxygen (aerobic) use a three-step breathing process. In the first phase, some organic molecules (glucose, fatty acids, and some amino acids) are transformed into molecular fragments with two carbon atoms (acetyl group). In the second phase, acetyl groups enter the Krebs cycle that oxidizes them into carbon dioxide and releases the energy that is stored in specific molecules. In the third phase, electrons and protons are released and electrons are transferred to oxygen through a series of electron-carrying molecules. Some intermediates that function as precursors of other metabolic pathways of this cycle of reactions may abandon the Krebs cycle. The citric acid produced in this cycle represents the metabolic crossroads on which converge both the degradative and biosynthetic pathways.

The discovery that a large number of biological processes involve cycles put cyclical organization (self-regulating cyclical processes with more or less sophisticated forms of feedback) at the center of contemporary biology. These processes include various biochemical metabolic cycles, cell cycles, reproduction cycles, and cycles through the biosphere, such as the carbon and nitrogen cycles. Therefore, cycles appear in biology at various levels of organization, and the best known of these cycles concern the relationship between organisms and their environment, an example of which is the nitrogen cycle. So far, we have focused on cycles at the level of basic metabolic processes. The presence of these cycles of reactions and their detailed knowledge allowed scientists to extract an organized system of the main pathways of cellular processes from what seemed to be a chaos of isolated reactions. They showed that some cyclical pathways follow a similar course in all living organisms. To use Krebs's words from his 1953 Nobel Prize acceptance speech: "We have long been familiar with the fact that the basic constituents of living matter, such as the amino acids and sugars, are essentially the same in all types of life. The study of intermediary metabolism shows that the basic metabolic processes, in particular those providing energy, and those leading to the synthesis of cell constituents are also shared by all forms of life."[12]

---

[12] Krebs, *The Citric Acid Cycle: Nobel Lecture*, 409.

Although we have already stressed this philosophical point, we wish to emphasize that the importance of self-regulating cycles in general and of metabolic cycles in particular is essential for overcoming the reductionist perspective. In a self-regulating cycle of reaction, there is no linear causality, which would be essential for epistemologically reducing the comprehension of a global process to an analytical understanding of individual processes. On the other hand, the presence of top-down causality, bottom-up causality, and cycle closure is essential for understanding these processes from a systemic point of view.

After discussing molecular chemical reactions, we will now consider models that permit schematization of the reactions taking place on macromolecules, focusing on the example of enzyme reactions.

Two Hypotheses of Enzyme Action

The glove-and-hand model, also called *induced fit*, and the lock-and-key model of enzyme reaction are two models showing how a molecule can bind to the active site of an enzyme. The lock-and-key model suggests that the molecular substrate has a complementary form to the active site, so that the two molecules adapt "perfectly," in the same way that a key (the substrate) fits into a lock (the enzyme). In this model, there is no change in the shape of the enzyme active site, when the substrate binds.

The induced adaptation model is similar to the lock-and-key model, but it asserts that the substrate and active site are not completely complementary but are only similar. Like a glove (the active enzymatic site) and a hand (the substrate), the shapes of the two molecules are similar but not the same. When a hand enters a glove, the glove slightly changes shape around the hand to fit it perfectly. Similarly, the active site changes shape to securely bind the molecular substrate. At the moment, the induced adaptation model is supported by many experimental results. For example, some enzymes can catalyze reactions with more than one substrate, and these different substrates have a similar shape in the same way that a glove can adapt to different hands.

The theory of induced adaptation explains a number of abnormal properties of enzymes. One such property is *noncompetitive inhibition*, in which a compound inhibits the reaction of an enzyme but does not prevent the binding of the substrate. In this case, the site where the inhibitor binds to the enzyme is not the active site but a site called the allosteric site, to distinguish it from an active enzyme site. The inhibitor changes the shape of the active site indirectly, through the changes of the allosteric site related to it. Allosteric sites are actually regulatory sites that can activate or inhibit enzyme activity by affecting the form of the enzyme so that, when the activator or the inhibitor dissociates, the enzyme returns to its normal form. In practice, enzyme flexibility is extremely important because it provides a mechanism to regulate enzyme activity.

Allosteric control can work in many ways. Let's examine the following example to illustrate a general effect. A pathway consisting of some enzymes is involved in the synthesis of the amino acid histidine. When a cell contains enough histidine, the synthesis stops. The synthesis is interrupted by the inhibition that the same histidine exerts on the first enzyme of the metabolic pathway. The inhibition of an enzyme by its own product is called feedback inhibition and occurs in many living processes. These processes are, therefore, also cyclical. Allosteric control can also be obtained through activators that act on the production of intracellular energy. A combination of allosteric activation and inhibition, therefore, allows the production of energy or materials when they are needed and interrupts their production when their quantity is sufficient.

Allosteric control is a quick method to regulate over time the products needed by living organisms. When the enzyme is not needed, some molecules called repressors, prevent its synthesis. Repressors are proteins that bind to DNA and prevent the first phase of the process leading to protein synthesis. When some substances that are added to cells are in need of an enzyme, an induced synthesis takes place. In the previous chapter, we saw how the presence of lactose or glucose, with its induction of enzyme synthesis, led to Monod's concept of the *operon*. Different cell types of multicellular organisms have different enzymes, although all cells in that organism contain the same DNA. The enzymes actually synthesized are those needed in a specific type of cell, and they vary not only for different types of cells (such as nervous, muscular, ocular, and cutaneous cells) but also for temporal necessity. Finally, if an enzyme consists of several subunits, the alteration of the shape of one of these subunits changes the activity in the others. As a result, the bond of a second substrate molecule occurs differently from the bond of the first, and the bond of the third occurs differently from the bond of the second. This phenomenon, called *cooperativity*, can favor successive interactions or it can disadvantage them. The first example of cooperativity was found in hemoglobin, which is a macromolecule with four similar subunits.

In completing this discussion of enzymatic activity, we return to the general consideration regarding ionic channels. To explain the enzyme activity, both proposed models focused on the shape of the macromolecule. We wish to emphasize that the shape is important, but so is the qualitative differentiation both of the active site and of other parts of the macromolecule. Parts of the macromolecule "direct" the molecule to the active site. There are other parts that can activate the same initial molecule and that can give it the right orientation.

All of this is visually highlighted when a macromolecule is brought back on a sheet of paper and on a screen. In this graphical representation, some chemical details are evidenced, particularly chemical groups, while other parts of the molecule are only hinted at. It is often said that some parts of the macromolecule are only there to orient the most active chemical parts. All of this describes the chemical approach to macromolecules, and its qualitative, as well as quantitative, diversity is what gives it richness compared to the purely physical approach.

The Energy Metabolism

We have already examined the division of metabolic processes into anabolic and catabolic types. The anabolic types allow the formation of new substances essential to living organisms, with their possible accumulation in cells as reserve substances. The catabolic types are processes of cleavage and degradation through which the cellular constituents and reserve substances are transformed into smaller molecules. To activate chemical reactions in general, we need a certain amount of activation energy. Moreover, each chemical reaction has its own energy balance given by the difference in energy between products and reagents. In general, the energy used by living organisms in chemical reactions is contained in molecules and is called chemical energy. In the course of the catabolic processes in particular, there is a liberation of this energy. This energy is in part released into the environment in the form of heat or work and is used in part to support the reactions that lead to the formation of other molecules. Therefore, each stage of metabolism must be examined in terms of both chemical and energy transformations. Often, the biological meaning of some chemical compounds that intervene in metabolic transformations is more connected to the energetic phenomena that take place as a result of such transformations than it is to the substances involved.

The set of energy transformations that occur in the various stages of metabolism is defined as energy metabolism, and it varies greatly depending on the activity that the living organism is carrying out. In the case of human beings, the basal metabolism is defined as that which is determined under conditions of fasting, immobility, horizontal position of the body, and so on.

From an energy point of view, living organisms are divided into two groups: autotrophic and heterotrophic. The autotrophic group processes its biological materials using directly absorbed solar energy. This group typically includes green vegetables that use chlorophyll to absorb solar energy. The heterotrophic group processes its biological materials by transforming the molecules taken in through food and from these molecules derives the energetic component of metabolism. This group typically includes animals. From the energy point of view, the chemical constituents of food for the heterotrophic group are real biological fuels. Heterotrophic organisms are constantly consuming energy that must constantly be replenished. The nutritional activity of these organisms provides this energy.

Chemical energy is supplied by organic substances that are transformed into carbon dioxide and water in the presence of oxygen, as well as by the organic molecules assimilated from the environment, mainly carbohydrates (sugars), proteins, and fats that are metabolized in the presence of oxygen. This process, which takes place through hundreds of intermediate chemical reactions, involves multiple energy transformations and the synthesis of intracellular reserve materials. At the end of the so-called intermediate metabolism, about one-third of the energy contained in the starting materials is made available to the cells. The remaining two-thirds are released and used during a subsequent series of metabolic reactions (terminal metabolism) that are often cyclical, such as the Krebs cycle that was previously discussed.

Two philosophical conclusions can be drawn from this discussion of energy metabolism. The first is that, although it is possible to distinguish a chemical from an energetic metabolism, the energetic metabolism can also be described from a chemical perspective, due to its formation, its accumulation, and its use. The importance of the molecule ATP (adenosine triphosphate) in energy metabolism seems to be a clear indication that the energetic processes occurring in living beings can be represented in chemical terms. The second philosophical conclusion relates directly to the systemic approach that we are proposing, and, to illustrate this conclusion, we will take the example of human nutrition as it relates specifically to the problem of obesity.

Obesity is a disease characterized by a pathological accumulation of body fat, with important consequences for a person's health and quality of life. Obesity currently constitutes one of the greatest public health crises in wealthy countries and has a decisive impact on life expectancy because it can lead to the onset of many diseases, such as diabetes and cardiovascular diseases.

In general, obesity is conceived as an *imbalance* between the energy that is consumed and the energy that is expended, this imbalance leading to the accumulation of excessive body fat. From this point of view, the two main factors relevant to obesity are nutrition and physical activity. A diet containing excessive calories is one factor that contributes to obesity, and a sedentary lifestyle is the other factor in the development of this condition, since the excess calories that are consumed are not expended through exercise and other activities. We have established that the human organism uses substances, primarily body fat, as an energy reserve. There is, therefore, a direct correlation between obesity and the energy and chemical metabolisms we have discussed.

From a systemic perspective, however, the problem with this mechanistic and reductionist approach to obesity is that this condition is mistakenly conceived only as a problem of imbalanced energy input and output. Yet as we have established, a living organism is an open system that under proper (homeostatic) working conditions can perfectly manage the variations of entry and exit. Thus, if the problem is the introduction of too much food into the body compared to energy needs, an organism with a properly functioning metabolism would have no difficulty in increasing the waste that it eliminates. Therefore, the problem of obesity and of all other metabolic pathologies is not primarily one of energy inputs and outputs. Rather, the problem lies in the organism's nonoptimal management of such inputs and outputs and, perhaps too, in the fact that such an open system was genetically calibrated in evolutionary conditions that were very different from the conditions of modern life in wealthy nations.

14

What Is Life? The Chemical Perspective and Its Relation to Other Perspectives

The title of this chapter refers to Erwin Schrödinger's famous 1944 book *What Is Life? The Physical Aspect of the Living Cell.* Schrödinger was an influential physicist and the co-founder of quantum mechanics, and his book had a considerable impact on developments in biochemistry during the second half of the 20th century, as well as on the nascent science of molecular biology. In the foreword to the most recent edition of Schrödinger's book, the 2020 Nobel Prize-winning physicist Roger Penrose provides evidence of its significance when he states:

> When I was a young mathematics student in the early 1950s I did not read a great deal, but what I did read—at least if I completed the book—was usually by Erwin Schrodinger. I always found his writing to be compelling, and there was an excitement of discovery, with the prospect of gaining some genuinely new understanding about this mysterious world in which we live. None of his writings possesses more of this quality than his short classic *What Is Life?*— which, as I now realize, must surely rank among the most influential of scientific writings in this century. It represents a powerful attempt to comprehend some of the genuine mysteries of life, made by a physicist whose own deep insights had done so much to change the way in which we understand what the world is made of.[1]

We refer to Schrödinger's important book for many and varied reasons that are all closely connected with our dominant theme in this work. We will focus here on two of these reasons, the first of which concerns the relevance of the question Schrödinger posed in the book's title. The second reason concerns the perspective Schrödinger adopted to address this perennial question. We examine whether the physical perspective that is advocated by Schrödinger is indeed the most appropriate for understanding the nature of life or whether the chemical perspective that we are advocating would be more suitable for this task.

The last few chapters of this book have traced the evolution of the chemical explanation of life. In this chapter, we try both to clarify what this perspective

---

[1] Schrödinger, *What Is Life? The Physical Aspect of the Living Cell*, 1.

From the Atom to Living Systems. Marina Paola Banchetti-Robino and Giovanni Villani, Oxford University Press.
© Oxford University Press 2023. DOI: 10.1093/oso/9780197598900.003.0015

means and why it is so effective and to examine the current state of affairs regarding the scientific understanding of life. We will also discuss the relationship between our systemic chemical perspective and other common perspectives, such as the processual and biosemiotic approaches. We will also schematize the chemical perspective and extend this discussion into the conclusion, where we will highlight the benefits of systemic explanations for the types of phenomena studied by the human and social sciences.

At this point, we will discuss Schrödinger's question in his book *What Is Life?* and will establish a link between this question and the relationship between chemistry and physics. After this, we will relate these two disciplines to the explanation of the realm of living beings, and we will cite the eminent 2020 Penrose again in this regard to demonstrate how this question is still relevant in our time.

At the beginning of *What Is Life*, Schrödinger identifies the purpose of his book as addressing the following question: "How can the events *in space and time* which take place within the spatial boundary of a living organism be accounted for by physics and chemistry?" This question has two fundamental aspects, both of which are related to the theme of our own book. The first and positive aspect is the question of how chemistry and physics can explain what happens in the "spatial and temporal boundaries" of a living organism. The second and, from our point of view, negative aspect is the coupling of physics and chemistry in this question. This coupling is common in the literature and presupposes the metascientific idea that these two disciplines occupy one and the same scientific and philosophical field.

If we add to this assumption the aggravating fact that the title of Schrödinger's book speaks of "the physical aspect of the living cell," with no reference made to chemistry, then Schrödinger's pairing of physics and chemistry is clearly meant in the reductionist fashion discussed in our introduction, according to which chemistry can be reduced to and absorbed by physics. It should not be forgotten that, in the first half of the 20th century, physicists believed that quantum mechanics could lead to the ontological and epistemological reduction of chemistry to physics. Paul Dirac, Schrödinger's colleague and co-founder of quantum mechanics, expressed this view clearly in 1929: "The underlying physical laws ... the whole of chemistry are thus completely known, and the difficulty is only that the exact application of these laws leads to equations much too complicated to be soluble."[2]

As has been established at length, one of the central themes of our book is precisely, on the one hand, the relationship between chemistry and physics and, on the other hand, the relationship between these two disciplines and biology. Thus, before considering in detail the chemical approach to the living, we

---

[2] Dirac, "Quantum Mechanics of Many-Electron Systems," 714.

reiterate our general position regarding how such relationships must be framed. From our philosophical and our systemic perspectives, we propose that the idea of a single scientific discipline to which all other disciplines must be reduced is an outdated notion because all scientific disciplines are parts of a whole, within which each has autonomy, in addition to being involved in interactions with the other disciplines. Thus, every scientific field, every disciplinary reality, has an autonomy that must be specified, but it also integrates with other scientific fields and disciplinary realities.

These interdisciplinary interactions are not external philosophical or scientific constraints, as the reductionist would claim. They are fundamental aspects of the scientific disciplines themselves, and they are as important as is the autonomy of each discipline. The scientific disciplinary system, like other systems, is concomitantly both closed and open. For this reason, we reject all disciplinary reductionism, not simply the reduction of chemistry to physics. Thus, while we demand the autonomy of chemistry and recognize an intrinsic cultural value in this discipline, we also demand the autonomy of biology and psychology and reject any attempt to reduce either of these disciplines. We would also extend this by saying that we demand the autonomy of the human and social sciences as well and reject any attempt to reduce them to the natural sciences. We completely dissociate ourselves from the physicalist thesis so narrowly characterized by the 20th-century philosopher Willard Van Orman Quine, when he states: "Physics investigates the essential nature of the world, and biology describes a local bump. Psychology, human psychology, describes a bump on a bump."[3]

As Schrödinger tells us openly, his scientific problem is that, although chemistry and physics can only explain biological phenomena with statistical laws, biology requires individual explanations. In addition, the quantum cohesions of the microscopic world are destroyed by the heat motion. Schrödinger states, "All the physical and chemical laws that are known to play an important part in the life of organisms are of this statistical kind; any other kind of lawfulness and orderliness that one might think of is being perpetually disturbed and made inoperative by the unceasing heat motion of the atoms."[4]

In response to Schrödinger, we emphasize that chemistry does not primarily explain through the use of laws, whether statistical or otherwise. As we have discussed throughout this book, the type of explanations of living systems offered by physics are distinct from the type of explanations offered by chemistry. Physics uses predominantly, though not exclusively, the concept of the laws of nature in order to explain how a single occurrence or phenomenon is the instantiation of a *universal* principle. Chemistry uses explanation by composition,

---

[3] Quine, *Theories and Things*, 16.
[4] Schrödinger, *What Is Life? The Physical Aspect of the Living Cell*, 10.

moving primarily horizontally from molecular complexity to the macromolecular complexity of the living being. Thus, chemistry identifies the entities that are present in the organism or in the biological process, and, through their previously identified static and dynamic properties, it reconnects the specific case to the *general*.

Although we will address the problem of the coherence within the microscopic world and of their "propagation" in the macroscopic world, we will first address the thermodynamics of nonequilibrium. Both the physical and chemical explanations represent two general modes of scientific explanation that we have elsewhere linked to the concepts of *universal* and *general*.[5] A single, unique fact must always be connected to other facts in order to be understood. The connection is always through the inclusion of individuals in "groups," whether these groups are defined by laws (all A's are B's) or, as in chemistry, by identifying collective properties. Grouping individuals into classes does not nullify their individuality, but it allows us to understand the properties common to several individual entities.

This dual physical and chemical approach is also used in the study of human behavior. Again, every action is unique and is performed by a single subject. However, if we stopped at this notion, we would not be able to connect different human behaviors, either individual or collective. Why did a person do one thing and not another? Why did such a historical fact occur? We can only come to understand individual and social actions by seeking general laws of behavior (the physical approach) or by grouping individuals into structured sets and using the static and dynamic properties of these sets, in a fashion similar to the chemical approach to explanation. Thus, although chemistry has always used such an explanation "through entities," it can offer a much more appropriate model of explanation than physics for the human and social sciences.

The book's conclusion will discuss this issue in greater detail. For now, we will examine more closely the effectiveness of these two types of explanation for understanding living beings and their characteristics.

The Efficacy of the Chemical Approach to Life

In addition to the specificities of different scientific explanations, one question stands out: Why is biochemistry more diffused and more efficient than biophysics for describing the realm of living organisms? With regard to the greater diffusion of biochemistry compared to biophysics, a search on the internet will yield a six to one ratio of results in favor of biochemistry. Although this is obviously not a

[5] Villani, *Complesso e organizzato. Sistemi strutturati in fisica, chimica, biologia ed oltre.*

scientific approach to measuring the diffusion of biochemistry over biophysics, it provides a clue to the preponderance of interest in biochemistry. The second part of the question, which regards the effectiveness of biochemistry in studying many of the processes within living beings, highlights the success of the chemical approach relative to the physical approach for understanding such processes.

The essay "La chimica: una scienza della complessità sistemica *ante litteram*,"[6] defined chemistry as an intrinsically systemic science, and, in our opinion, it is this approach that makes chemistry so efficient for describing complex realms such as that of living organisms. The chemical type of microscopic explanation, which deals with "pieces of structured matter" such as molecules and macromolecules and which highlights the static and dynamic properties of chemical entities and of their transformations, is more effective than physical explanations that search for universal laws for studying the complex world of a living being.

Let us clarify this a bit further by examining the following example, which was also analyzed in greater detail in *La chiave del mondo. Dalla filosofia alla scienza: l'onnipotenza delle molecole.*[7] Neurobiologist Pietro Calissano has explained why molecules are particularly suitable for studying the brain. His explanation is similar to the one generally developed in this book to explain the processes within living organisms. In speaking of neurotransmitters, Calissano states that

[they] play an essential role in communication between neurons. In the language formed by two fundamental symbols, the action potential and the neurotransmitter, the first can vary only in terms of frequency while the second has a much more articulated signal repertoire. Let's imagine that we have many rails on which vehicles run all the same. At the end of each rail there is a receptacle in which balls of different weight, colour or size are enclosed: each receptacle opens when touched by the incoming vehicle. The vehicle represents the action potential while the neurotransmitters are the spheres. The frequency of opening of each receptacle depends directly on the rate at which it is touched. It is evident that while the component constituted by the vehicles can vary only for the number of hits inflicted, the one formed by the receptacles is more articulated and versatile the more numerous and of different nature are the spheres. . . . It is not surprising then that the neurobiologists have identified first the system of the vehicles, that is, the action potentials and that only later have come to discover the neurotransmitters. It is even less surprising that while

[6] Villani, "La chimica: una scienza della complessità sistemica *ante litteram*."
[7] Villani, *La chiave del mondo. Dalla filosofia alla scienza: l'onnipotenza delle molecole.*

the fundamental properties of the former are now perfectly known, the real number of neurotransmitters used by neurons is still underestimated.[8]

To this day, 30 years after the publication of Calissano's book, the number of neurotransmitters is still underestimated. Obviously, the description involving the action potential, or electrostatic potential, that connects the neuronal network can be included in the physical approach to explanation, while the description involving chemical neurotransmitters belongs within the chemical approach. These specific chemicals are released by the first neuron at the synaptic interface between two neurons, and they activate the passage of a message to the second neuron. The greater versatility of Calissano's chemical explanation is evident to neurobiologist and is even more evident to us, once we understand that the traditional image of molecules with spheres of different color and size is a largely reductive one, especially when it is compared to the concept of qualitative differences in molecules and macromolecules.

In order to clarify the effectiveness of biochemistry for explaining biological processes, we will analyze another particularly interesting example, which is that of the immune system. Any standard biochemistry textbook[9] will explain the immune system by first defining what is meant by immune system. Thus,

[h]istorically, immunity meant protection from disease and, more specifically, infectious disease. The cells and molecules responsible for immunity constitute the immune system, and their collective and coordinated response to the introduction of foreign substances is called the immune response [and then it is completed by the action carried out by that system] . . . a more inclusive definition of the immune response is a reaction to components of microbes as well as to macromolecules, such as proteins and polysaccharides, and small chemicals that are recognized as foreign, regardless of the physiologic or pathologic consequence of such a reaction.[10]

Abbas, Lichtman, and Pillai go on to add that "innate immunity (also called natural or native immunity) provides the early line of defense against microbes. It consists of cellular and biochemical defense mechanisms that are in place even before infection and are poised to respond rapidly to infections."[11] And, finally

---

[8] Calissano, *Neuroni. Mente ed evoluzione*, 59–60.
[9] For one example of one of the best known textbooks in biochemistry, see Abbas, Lichtman, and Pillai, *Cellular and Molecular Immunology*.
[10] Ibid., 1–3.
[11] Ibid.

Because this form of immunity develops as a response to infection and adapts to the infection, it is called adaptive immunity. The defining characteristics of adaptive immunity are exquisite specificity for distinct molecules and an ability to "remember" and respond more vigorously to repeated exposures to the same microbe. The adaptive immune system is able to recognize and react to a large number of microbial and non-microbial substances. In addition, it has an extraordinary capacity to distinguish between different, even closely related, microbes and molecules, and for this reason it is also called specific immunity.[12]

The role played by molecules is already evident at this level, and, consequently, so is the role played by chemical explanations of this fundamental biological process, a role that was already highlighted in the title of the cited textbook in which the terms *cellular* and *molecular* are placed next to each other. Basically, "Innate and adaptive immune responses are components of an integrated system of host defense in which numerous cells and molecules function cooperatively."[13] Going into a little more detail, the textbook authors state that

[t]he main components of adaptive immunity are cells called lymphocytes and their secreted products, such as antibodies. Foreign substances that induce specific immune responses or are recognized by lymphocytes or antibodies are called antigens. . . . The modern definition of antigens includes substances that bind to specific lymphocyte receptors, whether or not they stimulate immune responses. According to strict definitions, substances that stimulate immune responses are called immunogens.[14]

From the repeated coupling of the terms *cellular* and *molecular*, two clearly separable aspects of the description of the phenomenon stand out. One is biological or cellular and the other is chemical or molecular. However, these are not in fact separable aspects. Abbas, Lichtmann, and Pillai state:

Since the 1960s, there has been a remarkable transformation in our understanding of the immune system and its functions. Advances in cell culture techniques (including monoclonal antibody production), immunochemistry, recombinant DNA methodology, and x-ray crystallography and the creation of genetically altered animals (especially transgenic and knockout mice) have changed immunology from a largely descriptive science into one in which

[12] Ibid.
[13] Ibid.
[14] Ibid.

diverse immune phenomena can be explained in structural and biochemical terms.[15]

Moreover, even when discussing active cells in the immune system, we must always resort to a chemical approach in order to understand how these entities relate to other cells or to external substances. This is evident even when considering the following simple numerical issue: "It is estimated that the immune system of an individual can discriminate 10^7 to 10^9 distinct antigenic determinants."[16] This requires an equal repertoire of lymphocytes: "This ability of the lymphocyte repertoire to recognize a very large number of antigens is the result of variability in the structures of the antigen-binding sites of lymphocyte receptors for antigens, called diversity. In other words, there are many different clones of lymphocytes that differ in the structures of their antigen receptors and therefore in their specificity for antigens, contributing to a total repertoire that is extremely diverse."[17] Clearly, the biological diversity of leukocytes, which are cells, is of a strictly chemical nature and is related to the receptor sites on their membranes.

Recently, two other general nonreductionist approaches to the study of life have come to the fore, notably, biosemiotics and processual philosophy of biology. We will not discuss these approaches in detail here because their importance and their enormous differentiation would require much more space to do them justice and this would go beyond the scope of this book. However, given our proposal for a predominantly chemical, systemic, and nonreductionist perspective to understanding life, it is imperative that we at least examine the relationship of our proposal to these other nonreductionist approaches.

In order to properly discuss biosemiotics, we must begin with the work of Jacques Monod. For our discussion of the philosophical importance of process in biology, we will have to examine the general use of the concepts of entity and process in science.

The Birth of Biosemiotics

The application of the semiotic approach to biology arose from two different biological phenomena: (1) the transcription of DNA into RNA and the translation of RNA into proteins and (2) the allosteric phenomenon present in some proteins. The semiotic approach was applied to the phenomenon of transduction

---

[15] Ibid.
[16] Ibid., 6
[17] Ibid.

of an environmental signal by a cell in which, as we have seen, one process occurs outside the cell and the other inside the cell. This approach was extended to other biological phenomena and over time became a generalized approach. All the aforementioned fundamental biological processes in living beings are characterized by the fact that they occur in two different positions and moments. What ensues in one of these steps is chemically independent of what ensues in the other step. This feature can be highlighted by analyzing the allosteric phenomenon of a protein.

A protein is said to be allosteric if it has two sites that perform two different functions. The first site is present in all enzymatic proteins, and it is called the active site because it is the site in which the protein performs its biological activity. The second site, which is called allosteric, regulates the activity of the active site by modifying, under appropriate conditions, what can occur in the active site. In particular, when a specific molecule binds to the allosteric site, the active site is "inhibited"; that is, it loses its "activity." Conversely, the active site requires the binding of a specific molecule in the allosteric site to become active. In such proteins, therefore, we can schematize these processes with three ligands, which are the two specific molecules and the protein and the ligand in the allosteric site that turns on or off the reactivity of the active site. The reactivity of the active site is interaction between the second ligand and the protein.

The important point here is that what occurs chemically on one site is disconnected from what occurs chemically on the other site. and the action on each site is only indirectly linked through the whole protein. This chemical independence led Monod to talk about chemically *gratuitous* processes:

> Here, let us emphasize what is by far the most important implication of the foregoing: namely, that the cooperative or antagonistic interactions of the three ligands are *totally indirect. There are no actual interactions between the ligands themselves; all the interactions occur exclusively between the protein and each ligand separately.* Further on we shall return to this idea, the apparently indispensable key to an understanding of the origin and development of cybernetic systems in living beings.[18]

An analogy may help to clarify the chemical independence of the two processes involved in the allosteric effect. Therefore, let us consider the macroscopic example of a double-plate balance. We place an object on a plate, and this action changes the situation in a second plate, either by balancing or imbalancing the second plate relative to the first. In this case, as in the allosteric effect, the important thing to emphasize is that the actions in the two plates do

---

[18] Monod, *Chance and Necessity. An Essay on the Natural Philosophy of Modern Biology*, 70–71.

not depend on the chemical composition of the objects involved. That is, an object made of iron, weighing 1 kg and placed in the first plate, will have the same balancing or imbalancing effect on the second plate as an object made of silver and weighing 1 kg.

This macroscopic example shows that we can extrapolate an important but overlooked characteristic of the allosteric effect. That is, the chemical indifference of the two processes is due to the dependence of this effect on a physical property. In the case of the two-plate balance, it is the physical property of mass placed on the first plate that alters the second plate. In the allosteric phenomenon, it is the geometric and physical property of the allosteric site that indirectly modifies the geometric and physical property of the active site. Thus, in practice, the fact that different chemicals linked to the allosteric site can change the active site is not specifically unique to this phenomenon. Rather, this phenomenon can occur whenever a protein/molecule interaction involves geometric/physical properties. Thus, we can think of the way in which, when decoding odors, the same molecular site can "respond" to molecules that are chemically different but that have similar shapes.

The scientific discipline of biosemiotics was developed precisely from the concept of the chemical independence of various processes. Although we agree with biosemioticians who investigate biological phenomena according to approach, we disagree with their claim that this approach can be generalized to, and is coextensive with life.[19,20] Even so, "chemical independence" does not cancel the need to study chemically the two steps of the phenomenon and, if possible, to identify the nonchemical properties that bind them and to understand the action of such properties.

The concepts of entity and process (or transformation) have always been the binomial that helps explain the persistence and variability of the world around us, ranging from the inanimate to the animate and then to the human species and its creations. In philosophy, this binomial is usually represented by the concepts of being and becoming and has been a central concern since the beginning of the Western philosophical tradition. Here, we must stress that there is no absolute distinction between entity and transformation, neither from the philosophical perspective nor from the more properly scientific perspective.[21]

In his many writings on the philosophical problem of entities and processes as they relate specifically to chemistry, Giovanni Villani demonstrated that the

---

[19] Sebeok, "Biosemiotics: Its Roots, Proliferation, and Prospects."

[20] Kull, "Vegetative, Animal, and Cultural Semiosis: The Semiotic Threshold Zones."

[21] A more detailed discussion of entities and processes would take us far from the central topic of this chapter. We refer the interested reader to the following works by Giovanni Villani, which address this topic as it relates specifically to chemistry: Villani, "Ruolo degli enti e dei processi in chimica e in fisica"; Villani, Chemistry: A Systemic Complexity Science; Villani, Complesso e organizzato: Sistemi strutturati in fisica, chimica, biologia ed oltre.

components of a chemical whole all act together and dynamically toward the formation of a new system. Once the new system is formed, however, one must then consider its global action within the environment. Moving within the level of the macromolecule's complexity, the system's properties are global, but they are not necessarily static. As emphasized earlier in this book, the system is an intrinsically dynamic entity. Even when considering the component level of the system, we encounter properties that cannot be traced back to the individual parts. The fact that we must add a certain number of water molecules to describe the environment around the system and its interaction with the macromolecule does not change the terms of the systemic interaction. In fact, the system must always be described with the "internal" and specific characteristics of the macromolecule, as well as those of the water molecule and of the interaction of these structured entities.

As discussed in Chapter 11, the complexification of the proteome requires more, rather than less, chemical understanding, and, in particular, it requires a systemic approach to chemical understanding. Moreover, in cells, we must foreground sets of reactions (whether cyclical or otherwise), rather than individual reactions, and this enriches the systemic dynamic problems that our approach seeks to understand. However, before moving on to addressing the question "What is life?," we must first distinguish between different types of systems.

Open and Closed Systems

In order to examine different types of systems, we must first recognize a fundamental philosophical and scientific distinction between open and closed systems. A system is *isolated* when it does not exchange energy with its environment, and a system is *closed* when no material enters or leaves this system. Finally, a system is *open* when there is an exchange of energy and matter between it and its external environment. Since, according to modern physics, there is no absolute differentiation between the energy and the material particle that carries it, the well-defined thermodynamic difference between an isolated and a closed system becomes much smaller.

In addition, no system is totally isolated. Even thermodynamically isolated systems are "open" from the point of view of gravitational and electromagnetic interactions. At most, a totally isolated system that has no interaction with the outside environment would be a system about which it would not be possible to acquire information. On the other hand, thermodynamically open systems must have a closure that preserves the individuality of the system. That being said, the differentiation between open and closed systems remains a cornerstone of scientific analysis since the opening of energy and material, and thus of information and communication, is more than the interactive and relational openness of any given system.

Physics, and especially thermodynamics, has long studied closed systems almost exclusively, as well as the conditions that would bring them to a state of equilibrium. As already mentioned, open systems do not reach a state of equilibrium but instead something called a "stationary/steady state." The term *homeostasis* refers to the set of processes that act to maintain this state, both in its morphology and in its internal conditions, in spite of any external perturbations. In 1932, von Bertalanffy advanced the concept of the organism as an open system and followed this with the development of general kinetic principles and their biological implications.[22,23] However, it was the work of Iya Prigogine in the second half of the 20th century that defined and developed thermodynamic concepts as tools for understanding open systems.

In the context of studies on open systems, concepts such as feedback have contributed important distinctions and clarifications regarding causality. The field of cybernetics has also contributed significantly to this work by introducing information devices that provide feedback to identify and reverse changes. A fundamental postulate of cybernetics is that "everything is connected to everything else." Norbert Wiener, who founded the field of cybernetics, focused on systems in which A and B are concomitantly the cause and effect of each other. In such systems, the effect affects the cause through feedback, and this feedback creates a specific circular causality. Given this holistic connection, which is very distinct from the traditional approach of physics, the goal was to materialize a network of actions and feedbacks through which both living and nonliving complex and organized systems could adapt and survive in a changing environment. In addition, it was hypothesized that there were no substantial differences between living organisms and complex self-regulating machines, so that the behavior of both could be described by the same theory.

Von Bertalanffy's concept of *equifinality* is the concept of multiple paths that lead to a common end state. That is, some systems will reach the same endpoint, regardless of the starting point of each system. For example, the phenomena of life behave in this way, since the final state can be reached from very different initial conditions and in very different ways. Morin states: "Equifinality means that a system can, according to the alee, the difficulties, the resistances it encounters, use different strategies to reach the same goal, and that more similar systems can achieve the same ends with different means."[24]

[22] Bertalanffy, "Der Organismus als physikalisches System betrachtet," 521.
[23] Bertalanffy, "The Theory of Open Systems in Physics and Biology."
[24] Morin, *Il metodo 1. La natura della natura*, 312.

We should note that equifinality was long considered as primary evidence for *vitalism*. According to von Bertalanffy, closed systems cannot behave according to equifinality, which is why equifinality is generally found not in inanimate systems but, rather, in open systems that exchange materials with the environment. Because these systems reach a stationary state, that state is independent of the initial conditions and is considered equifinal. Von Bertalanffy adds, however, that equifinality is also found in some open inorganic systems so that these exhibit paradoxical behavior, as if these systems "know" the final state that they must eventually reach.

An important feature of many open systems that is not exclusive to living systems is the fact that equifinality occurs even in highly complex systems that, from an a priori perspective, could reach several possible endpoints. As an example, we can consider the native form of a protein, something we have already examined in detail. If analyzed in terms of potential energy, the protein has many possible stable forms. But in the reality of its actual environment, the protein manifests almost exclusively in one form. In sum, all open systems share the following common characteristics:

- The ability to not move toward maximum entropy
- Feedback
- Homeostasis
- A remarkable equifinality

The fact that entropy can decrease in a system, as the system increases its order, indicates that systems can survive and maintain their characteristic internal order only if they import more energy from the environment than they expend on their internal processes and on the export of their products. The feedback principle can be related to the information introduced into the system, which can be seen as a particular type of signal that the system and the environment can exchange and that provides the system with information about the environmental conditions and the functioning of the system itself. The feedback provided by this information allows the system to remedy a malfunction or to adapt to environmental changes and, thus, to maintain a steady state or homeostasis. The remarkable equifinality of open systems enables them to reach the same final state even when starting from very different initial conditions and when following very different lines of development. Finally, under appropriate conditions, open systems move in the direction of differentiation and complexity in which global action models of a generic nature are replaced by more specialized functions.

Thus, the sharp contrast between order and disorder that, for many centuries had been seen as an essential difference between the animate and the inanimate has been replaced by the contrast between open and closed systems.

Development and evolution, which are evident in living things through the emergence of higher orders and their differentiations, are not specific characteristics of living systems but of all open systems, and they always involve extracting energy from the environment surrounding those systems.

We have emphasized the usefulness of describing living beings as open systems that exchange matter and energy with their environment. However, it is not enough to demonstrate the thermodynamic that is implicit in that analysis. It is imperative that, in addition to thermodynamics, we also consider kinetics, as chemistry does in the study of transformations. In order for the reactions essential to life to take place, the energy barriers to triggering such reactions must be overcome and living beings require the acceleration of even relatively rapid reactions.

More than 99% of the reactions relevant to biological systems are catalyzed by enzymes. Therefore, they are fundamentally necessary for the existence of life. Life as we know it would not be possible without enzymatic catalysts. However, since we have already discussed enzymes, we must now examine in some detail the thermodynamics of the nonequilibrium of open systems, which was elaborated primarily by Prigogine and which is essential for explaining life.

Prigogine and the Thermodynamics of Nonequilibrium

We have seen that, from a strictly thermodynamic point of view, one can make the following distinction between open and closed systems. Closed systems display equilibrium where the forces are balanced, and, in this equilibrium state, the system contains the greatest possible amount of entropy that is compatible with the particular conditions in which it exists. On the other hand, open systems display dynamic steady states, which are maintained by a continuous expenditure of energy.

Open systems display an interrupted influx of energy from the external environment and a continuous outflow of products. Despite this, the character of an open system, the relations between its different components, and its relationships of exchange with the outside environment do not change. For example, catabolic and anabolic metabolic processes are necessary for living beings to generate a stationary state of the cell, but the effect of these processes on the chemical "content" of cells is not always the same. However, this content is very similar to that of the previous state, and the living being can continue performing all of its chemical transformations.

The Second Law of Thermodynamics introduces the idea of the degradation of energy, rather than the idea of loss, which would contradict the first principle. While all other forms of energy can be transformed integrally into one another, the form of energy that we call heat cannot completely be converted and therefore makes a system lose some of its ability to perform a task. This irreversibly decreased ability to transform and perform a task, which is proper to heat, is what Clausius called entropy. According to this principle, irreversible degradation can only grow to a maximum, which is the state of homogenization and thermal equilibrium.

The principle of energy degradation as formulated by Clausius along with Sadi Carnot and William Thomson, Lord Kelvin was later called the principle of order degradation by Boltzmann, Gibbs, and Planck. Thus, the concept of entropy refers concomitantly to the degradation of energy, order, and organization. Although the Second Law of Thermodynamics remains valid for the open system and its environment, it does not prove to be at all valid for the open system itself.

The importance of the Second Law of Thermodynamics for closed and open systems can be expressed in a different way. This principle asserts that the general tendency is oriented toward states of maximum disorder, that the higher forms of energy (mechanical, chemical, and luminous energy) are irreversibly degraded to thermal energy, and that thermal gradients must disappear. From a philosophical perspective, the Second Law of Thermodynamics stands out as the one universal law of the cosmos, which applies not only to all physical objects as conceived in isolation but also to the beginning and ending of the universe itself. However, in this perspective, there arises the problem of understanding why and how order and organization exist at all. To the extent that the Second Law of Thermodynamics is for closed systems, it is valid for a whole open system plus its environment, since these together constitute a closed system.

For the open system considered in isolation, which does not abide by the Second Law of Thermodynamics, "degradation" is not necessary. Thus, one can create local order and cannot locally degrade the best forms of energy while maintaining thermal gradients. Therefore, and as mentioned earlier, the sharp contrast between order and disorder must no longer be considered as distinguishing animate and inanimate nature but, rather, as distinguishing open and closed systems. In addition, the development and evolution evident in living beings are not simply characteristics of life but of all open systems. Under appropriate conditions of competition, all open systems will move in the direction of differentiation and complexity, in which global schemes of generic action are replaced by more specialized functions.

We have established that 19th-century scientists found it difficult to understand how to relate the Second Law of Thermodynamics to the order of both the living individual and the evolution of the species. Let us now examine how contemporary science responds to this problem. The Second Law of Thermodynamics does not preclude the parts of a subsystem within a larger system from transforming over time toward a more ordered state, thereby enabling that system to evolve into a more organized and complex form.

The Second Law of Thermodynamics simply states that, in such transformations, the total entropy of the isolated system always increases. Thus, if the creation of life and order implies a dynamic transformation toward a structured and organized state, this transformation is inevitably accompanied by a greater increase in disorder in the (almost) isolated system that includes the Sun and the Earth. In other words, although some parts of this isolated system will gain entropy and others will lose it, the total entropy of this isolated system will always increase over time.

Our Sun is the predominant source of energy that sustains life on Earth. All organisms maintain a highly organized state by extracting solar energy from the surrounding environment and exporting waste into it, thus producing entropy. Importantly, the Sun continuously provides the Earth with high-quality (i.e., low-entropy) free energy that keeps Earth's entropy low. The low entropy of the Earth is then primarily maintained through phototrophic organisms (i.e., plants, algae, cyanobacteria, and all photosynthesizers) and chemotrophs (e.g., iron and manganese oxidizing bacteria found in igneous lava rock), which are the first organisms to use the free energy of solar photons and inorganic compounds.

For example, most green plants have a relatively large surface area per volume unit that allows them to capture a high amount of free energy from the Sun. Specifically, the absorption of photons by photosynthesizers generates paths for the conversion of energy, coupled with photochemical, chemical, and electrochemical reactions that lead to high-quality energy. This energy is converted into useful tasks that keep such living beings in a complex, high-energy, organized state that, in turn, supports all the heterotrophic (animal) life on Earth.

Prigogine seeks to explain the principles governing the relative stability of ordered and, sometimes, highly ordered systems in a universe that is governed by the Second Law of Thermodynamics. He achieves this through the concept of a dissipative system, of which the biosphere of planet Earth serves as the best example. A dissipative system is one in which the continuous flow of energy and matter coming from outside permanently keeps that system from achieving thermodynamic equilibrium, so that it can indulge in the creation of negentropy (or negative entropy) in the form of ordered structures and living organisms.

An important aspect of Prigogine's theory concerns the fluctuations of systems from equilibrium because, under specific conditions, they can be considered the engine of evolution in nature. Through a progressive departure from a thermodynamic equilibrium, such systems determine a spontaneous and stable growth of organization. It is well known that the state of thermodynamic equilibrium does not possess those characteristics of absolute homogeneity that are predicted by the theory, but only possesses these at a statistical level. The extensive quantities are distributed within a system with a density variable from point to point and from instant to instant around the mean value. Consequently, the values that should be attributed to the intensive quantities are also subject to local fluctuations. In closed systems, small fluctuations are enough to cause intense recall forces that cancel these out. On the other hand, under specific conditions, recall forces are generated only when the fluctuations are large enough. In such cases, small fluctuations can grow undisturbed and, therefore, become extremely intense near critical points.

As a result of fluctuations, the Second Law of Thermodynamics displays a probabilistic character only. For example, an exchange of heat can also take place from a system at a lower average temperature to a system at a higher average temperature. But this exchange is less likely in the reverse direction, so the Second Law of Thermodynamics will be respected in the long run. In other words, the laws that govern thermodynamic processes are deterministic as long as one is dealing with macroscopic exchanged quantities, while infinitesimal exchanges can only be described probabilistically.

This situation becomes even more complex once one considers situations of nonequilibrium. Prigogine analyzes a particular case of this type of situation, the Bénard instability that occurs in a liquid layer that is heated from below.[25] Once a certain threshold value has been exceeded, the system creates convention currents that result from the nonequilibrium interaction between the heat flow and gravitation. Such convention currents include a number of molecules on the order of 10^{21} in size, which is an extremely large number of particles. Therefore, the nonequilibrium has created coherencies between very large numbers of molecules, which allows the particles to interact over long distances.

Using a beautiful and incisive metaphor, Prigogine states that he likes to think of matter in the vicinity of equilibrium as "blind" because each particle only "sees" the molecules that immediately surround it. On the other hand, when far from equilibrium, long-range correlations are produced that allow the construction of coherent states, which are studied today by numerous branches of physics and chemistry. The example of Bénard's instability is not an isolated case. In recent years, the appearance of nonequilibrium structures has also been observed in different areas of hydrodynamics and, primarily, in chemistry. The conditions that lead to nonequilibrium structures can be generated as soon as autocatalytic phenomena exist, that is, as soon as one can obtain feedback that amplifies kinetic phenomena.

Prigogine's development of the thermodynamics of nonequilibrium demonstrates that exclusion is not necessary and that complementarity between disordered phenomena and organizing phenomena is possible. Deviance, perturbation, and dissipation can generate a "structure," that is, they can generate organization and order at the same time. This concept is captured in Heinz von

---

[25] Prigogine, *La nascita del tempo*, 70.

Foerster's "order from noise principle,"[26] in the organizing principle that emerges through Edgar Morin's concept of disorder,[27] and in Henri Atlan's organizing case.[28]

Morin goes even further than Prigogine when he states that every open and complex system is imaginable as a "conflicting whole." In such systems, there emerges a manifest unity and, at the same time, a latent antagonism, which is the potential bearer of disorganization and disintegration. Morin calls this feature "the principle of systemic antagonism." The idea of a system is not simply the idea of harmony, functionality, and a superior synthesis. This idea also carries within it the necessity of dissonance, opposition, and antagonism. Every system in which organization is active is actually a system in which antagonisms are also active. The organization tolerates a margin of fluctuations, which must be kept below a certain threshold to avoid disintegration. Thus, the active organization connects complementarity and antagonism in a complex and ambivalent way.

The growth of complexity within the organization corresponds to a new potential for disorganization. The disintegrating effect of antagonisms can only be counteracted through active resistance. This explains why no organized system is eternal. Eventually, internal antagonisms will always disintegrate a system. As an example, limited lifetime is a specific and distinctive feature of every living organism, and so the extent to which this approach is applicable to the human and social science is clearly evident.

System-Environment

In general, a system is delimited by the available surrounding environment with which it interacts and which conditions all of the system's actions. In particular, the inputs that enter an open system do not solely include energy materials that are intended to be transformed or modified by the tasks carried out in the system. These inputs also include information that is provided to the system's internal organization, "signals" about the environment in which the system is located, and information about the operation of the system in relation to that environment. If an organism is to be adaptable to a particular environment, its modifications do not only depend on the characteristics and nature of the organism, but also equally on the environment and the possibility for the organism to "know" that environment. We understand that this is true of living systems, but it is also a feature of all open systems.

---

[26] Foerster, "On Self-Organizing Systems and Their Environments."
[27] Morin, *Il metodo 1. La natura della natura*, 56.
[28] Atlan, "Sul rumore come principio di autoorganizzazione," 35.

In order to properly study a system, it is imperative to focus on specific properties to the exclusion of others and, therefore, to separate the system from its environment in some way. To be more precise, in the system/environment pair, what constitutes the inside and the outside of the system must be clarified according to the organizational order of that system and the particular process or properties being studied. Thus, the system and its environment may be conceptually divided in different ways, depending on the analysis being conducted. This means that the concept of environment, which originated within the field of biology, must be extended to the entire systemic approach.

Two extreme perspectives on the relationship between an organized system and its environment have been proposed. The first perspective, the ecological view, conceives of systems as being immersed within their environments. The second perspective considers the environment to be free of structure and to serve as a sort of "background noise." This perspective denies the presence of relations or specific interdependencies between the properties of the system and those of its environment. Thus, the choice of one of these perspectives when studying a system will alter how the role of the environment is conceived.

Unfortunately, this choice is often made in an implicit and unclear fashion and ultimately, it leads to a considerable difference in the way that scientists and philosophers will apply the systemic approach to study organized wholes. However, we believe that understanding the relationship between open and organized systems and their environments must always be included in such considerations because it is only through exchanges with their environment that living organisms can survive.

A Philosophical Consideration of *What Is Life* and Schrödinger's Problem

As has already been stated, Schrödinger postulates that there are two types of order: order-from-disorder and order-from-order. He states that the natural order, as described by the laws of physics, was based on atomic statistics and therefore was only approximate: "The laws of physics and chemistry are inaccurate within a probable relative error of the order of $1/\sqrt{n}$, where n is the number of molecules that co-operate to bring about that law—to produce its validity within such regions of space or time (or both) that matter, for some considerations or for some particular experiment."[29]

What is novel for the physicist with regard to living organisms is that their stability is the result of a process of order by order. For Schrödinger, this order

---

[29] Schrödinger, *What Is Life? The Physical Aspect of the Living Cell*, 17.

is due to the substrate of genetic material being an aperiodic crystal, that is, a structure with a specific but not periodic arrangement of atoms, encoding the information that guides the development of an organism., Therefore, life must acquire its capacity to evade the degrading forces of heat motion through what can be called a translation of the order of the aperiodic crystal to that of cell dynamics. The idea of a translation from the crystal code to cell chemistry would seem to follow specifically because the aperiodic crystal is protected and sequestered from the cellular chemical dynamics. Therefore, Schrödinger denies precisely the notion of dynamic self-organization resulting from statistical interactions, that is, the order that results from disorder under nonequilibrium conditions.

It would seem, however, that the organismal order may require a different type of explanation than the one provided by Schrödinger, since such order cannot be physically explained through DNA but only through the organism as a systemic whole. We now know that neither DNA nor any aperiodic crystal is a unique repository of hereditary stability in the cell. The structural/functional organization of the cell, the basic membrane system, and cell compartmentalization are all transmitted from one generation to another through the maternal egg cell. Following the categorizations given by Jablonka and Lamb,[30] these aspects are grouped into three epigenetic inheritance systems (EIS): (1) organizational structure, (2). steady-state dynamics, and (3). chromosome marking.

In addition to this biological response to the problem posed by Schrödinger, we want to make a further point. In Schrödinger's time, quantum chemistry was still in its infancy. However, with our current understanding of single and specific molecules and their molecular structure, we can explain the properties and aspects of the world around us, whether inanimate or animate. This is why we believe that chemistry is a science of systemic complexity and that it is through its approach, conjoined more broadly with systemics, that we can unify and explain the entire material world. In fact, the potential of the chemical approach to the study of life that we are proposing in this book are demonstrated precisely by the order of the living world and the problem posed by Schrödinger.

To conclude this final chapter, let us reexamine the original question, "What is life?, and ask whether it makes sense to ask this question today as it did in Schrödinger's time. Underlying this question are myriad more general problems that transcend the purely scientific meaning of life. For example, in his book *At Home in the Universe*, Stuart Kauffman wonders: "If I am right, the motto of life is not We the improbable, but We the expected"[31] because, by addressing questions

[30] Jablonka and Lamb, *Epigenetic Inheritance and Evolution.*
[31] Kauffman, *At Home in the Universe: The Search for Laws of Self-Organization and Complexity*, 51.

regarding life, Kauffman wishes to uncover the place of humans in the universe. These are important and pertinent questions, but they may not be answerable from a scientific perspective, and besides, they are beyond the scope of this book and have already been treated in depth in many other works. Instead, in closing this chapter, we wish to foreground an aspect of the question "What is life?" that is closely related to the theme of this book, in order to establish whether the systemic chemical approach that we advocate can shed some light on the possible answer, even if it is only a partial light.

Although many, and at times diametrically opposed, scientific answers have been proposed to this question, it is clear that no definitive resolution has been provided. In order to address the question "What is life?" scientifically, one must be able to identify a group of properties or phenomena that identify a being as alive and that are not found in nonliving beings. Such defining properties or phenomena can be called the "necessary and sufficient" conditions for establishing that a being is alive. There is yet no agreement on what these defining properties or phenomena are, but this is probably a minor consideration for us here.

The real difficulty for science lies in the question of what "life as such" is and which scientific discipline is best equipped to address this question. There is still disagreement regarding whether chemistry is capable of helping to resolve this problem, although it can certainly explain many or all processes that take place in living organisms. The leading 20th-century microbiologist Carl Richard Woese reflects this concern very clearly:

> Biology today is at a crossroad. The molecular paradigm, which so successfully guided the discipline throughout most of the 20th century, is no longer a reliable guide. Its vision of biology now realized, the molecular paradigm has run its course. Biology, therefore, has a choice to make, between the comfortable path of continuing to follow molecular biology's lead or the more invigorating one of seeking a new and inspiring vision of the living world, one that addresses the major problems in biology that 20th century biology, molecular biology, could not handle and, so, avoided. The former course, though highly productive, is certain to turn biology into an engineering discipline. The latter holds the promise of making biology an even more fundamental science, one that, along with physics, probes and defines the nature of reality.[32]

To address the question of what life is and to avoid philosophical or religious considerations, we might proceed by asking a scientifically related question, "What is molecular structure?," since the view advocated in this book does

---

[32] Woese, "A New Biology for a New Century."

not recognize any absolute dichotomies between living and nonliving entities or systems. The book's journey has taken us from the undifferentiated atom to distinctive chemical atoms, from atoms to molecules, and from molecules to macromolecules.

We have discussed, from an epistemological and nonmetaphysical perspective, the emergence of new properties involved in the transitions between levels of complexity and have established that each of the transitions, from less structured to more structured and from less complex to more complex, generates specific emerging properties. These emerging properties are due to the organization and structuring of the components that generate a new and global system. This is the case for the emergent property that we call molecular structure in the transition from the aggregate of atoms to the piece of organized reality that we call a molecule. This is also the case for the property that we call life, which emerges in the transition from the aggregate of chemicals in molecules and macromolecules to the cell system in which these are organized.

Reasoning in this manner, we do not think it is out of place to compare the organization of the molecular system to the organization of the cell and, ultimately, to the organization of the living system. To highlight an important aspect of the definition of such emerging entities, we will examine the structural and functional definitions of molecule and cell. From a structural perspective, a molecule is an entity composed of two or more atoms held together by some type of "glue" that is generally referred to as covalent bonds. From a functional perspective, such as that taken by Cannizzaro, a molecule is the smallest amount of chemical that enters in specific reactions.

Leaving aside the term *biological*, which is often encountered in the definition of the cell, we note that such a definition can also be either structural or functional. One might argue that the structural definition is reductionist, referring to the simplest complexity level: that of the constituents of a cell. One might also argue that the functional definition is holistic, referring to the complexity level in which the cell performs a function. In reality, however, such definitions connect these various levels of complexity of molecules and cells in a circularity. Both the molecule and the cell are defined in terms of what they are structurally and what they do functionally with respect to each other.

Within our systemic perspective, we believe that the internal or structural aspects and the external or functional aspects are complementary and are both necessary to the definition of organized system, a system that is concomitantly closed and open. Such a systemic perspective not only connects different levels of complexity but makes possible a diversity of types of explanations that are top-down, bottom-up, and circular. Banchetti-Robino has argued elsewhere that the chemical approach to understanding quantum systems is very effective for understanding what is considered one the of the most mysterious emergent properties of living systems, that is, consciousness.[33] Therefore, to close this chapter, we state that life is the emergent property that interconnects the chemical planes of molecules and macromolecules to the global functions that these grouped and organized components perform at the biological level. In our view, there is no new biology to be created that can surpass the molecular and the macromolecular levels because, in all cases, the biological explanation transcends these two levels and is directed at addressing questions that are not chemical but globally biological.

[33] Banchetti-Robino and Llored. "Reality Without Reification: Philosophy of Chemistry's Contribution to Philosophy of Mind."

Conclusion

As we arrive at the conclusion of our work, we will outline and analyze the systemic chemical explanation of the material world that we have defended throughout this book. Our historical path has carried us from the 17th-century conception of the undifferentiated atom to the 20th-century conception of chemical wholes as organized systems. We have characterized chemical explanation and have discussed its application to the wide fields of biochemistry and molecular biology. At this point, we will highlight the specific philosophical characteristics of our thesis to suggest that the application of the systemic approach used in chemistry serves as an explanatory model for both individual and social human phenomena.

We begin by discussing two philosophical problems that are implicit in the term *systemic*. The first is the problem of identifying and conceptualizing what we have called levels of complexity and their interrelations. More simply put, we must address the broad philosophical theme of the structuring of the world, a problem that can be traced as far back in time as the birth of Western philosophy. The second problem, which is much more recent, concerns downward (or top-down) causation. Both of these problems are of such philosophical breadth that our biggest challenge is to avoid trivializing them in the brief space available here for this discussion. However, once we have addressed these problems, we will be able to address the manner in which the systemic chemical model of explanation can serve as a model for the human and social sciences and their explanations of living complex systems.

Structuring the Physical World (and More)

Setting aside the question of whether the structuring of the world is correct, there are many ways of considering how this structuring can be done that do not require us to return to earlier philosophical perspectives. Here, we will focus on two of the many ways of structuring the world. The first is compositional and is sometimes called *layer-cake structuring*. The historical precedents for this approach are many, so, for the sake of simplicity, we will focus on the modern version of this structuring, which was promoted by Hilary Putnam and Paul Oppenheim.[1]

---

[1] Putnam and Oppenheim, "Unity of Science as a Working Hypothesis."

Layer-cake theory has three components. First, it claims to be a comprehensive theory that can account for all types of phenomena that can be organized according to levels. Second, the levels are correlated according to compositional relationships that are structured in a gradual, *step-like* manner. In practice, all entities present at one level are regarded as composites, and the entities at the adjacent lower level are viewed as the constituent parts of the composites at the immediately higher level, and so on for each level above the lowest and most fundamental level. This structuring of layers on top of layers gave this theory its name. Third, Putnam and Oppenheim assume a close correspondence between the entities that make up a level and the predicates and scientific theories related to them. This implies that the various levels studied by modern science neatly correspond to the levels of nature. Thus, for each level of nature there is a specific scientific discipline, with its own specific theories, that explains the phenomena at that particular level.

Putnam and Oppenheim propose six organizational levels. From lowest and simplest to highest and most complex, they are as follows: elementary particles, atoms, molecules, cells, multicellular living things, and social groups. Physics studies the levels of elementary particles and of atoms, chemistry studies the molecular level, and so on.

This structuring model can be regarded from either a reductionist or an antireductionist perspective. Putnam and Oppenheim adopted the reductionist perspective, which is a central element in their argument for the unity of science, not simply as a theoretical possibility but primarily as a working hypothesis. This is because they argued that the more specific laws of science within each higher level are themselves subsumed by the more general laws of science at the adjacent lower level, until it is concluded that the laws of elementary particle physics are the most general laws that subsume all of the more specific laws for higher-level phenomena. Putnam and Oppenheim go so far as to claim that some of the laws of chemistry have already been reduced to physics, and they expect that, ultimately, this will result in a complete reduction of chemistry and all the other higher-level sciences to physics.

From the antireductionist perspective, however, although the reality of levels is acknowledged, each level is regarded as fundamental in its own right, and its phenomena and properties are not considered to be completely reducible to the lower levels. This also entails that the laws and theories within a given scientific discipline are themselves not completely subsumable by the laws and theories of the disciplines that study lower levels. Apart from focusing on the number of levels, which in our view should include the macromolecular level, the systemic perspective alters the term *entity* to *system* and also considers the emergence of novel properties at each level, thereby generating the classic model of systemics.

There are many possible criticisms of the layer-cake model of structuring the world. One criticism acknowledges that scientific disciplines do not fall neatly into this scheme and that each science studies and works with multiple levels of complexity, with chemistry serving as an excellent example of this approach.

Given the many problems associated with the layer-cake model, we will now consider the second approach to structuring the world: the procedural model. Here too there are many possible examples of this model, so we will focus on the mechanism approach such as the one derived by neuroscientists Carl Craver[2] and William Bechtel.[3] This approach proposes a contextualized conception of the ontological levels of nature, defining these levels in terms of constitutive parts within a mechanism. The governing logic of this approach is that the processes of a mechanism prevail on the entities within the mechanism. Therefore, the levels are identified by the mechanisms that express them.

This procedural model shifts attention from material composition to process. Regardless of whether this is done through the concept of mechanism or in some other way, this approach has become widespread in the field of biology and has led to the conception of life itself as a process. According to this model, the relevant parts within a mechanism are working parts, that is, the parts that perform the functions that allow the mechanism to carry out a process. These parts may be of different sizes, but they all participate in the functioning of the mechanism. In this model, a level is constituted by the set of organized working parts whose operations are coordinated to achieve the process that is being studied. Obviously, the number of levels involved in any given process cannot be known in advance, so, in practice, there is no a priori limit to the number of levels in a mechanism.

This discussion of processes prevailing on their functional parts brings us to the many philosophical difficulties involved in accepting downward causation, which arise from the fact that the parts making up a system are simple entities that are assembled together. Thus, the idea that such parts could exert causal influence on the system itself seems incoherent. This problem has tacitly informed many of the perspectives in the history of modern science that we have discussed in this book, beginning with the mechanistic conception of atoms as simple entities that are "joined," broadly speaking, to form a molecule.

As we have pointed out, this idea reduces the molecule to an aggregate of atoms, the cell to an aggregate of molecules, the macromolecule to an aggregate of molecules, and so on. Thus, we agree with Charbel Niño El-Hani's claims that "[w]hen lower-level entities are composing a higher-level system, the set of possible relations among them is *constrained*, as the system causes its components to

[2] Craver, "Beyond Reduction: Mechanisms, Multifield Integration and the Unity of Neuroscience."
[3] Bechtel, *Mental Mechanisms: Philosophical Perspectives on Cognitive Neuroscience.*

have a much more ordered distribution in space–time than they would have in its absence This constraint on the components' relations results from their being part of the space–time form, or pattern, of the system's structures and processes."[4]

We grant that the notion that parts of a system are different from what those parts would be if they existed in isolation is somewhat controversial, but we are not alone in endorsing this view. John Dupré, for example, states: "My central claim is that the properties of constituents cannot themselves be fully understood without a characterization of the larger system of which they are part."[5] We argue that the explanation for this difference lies in the structuring of the parts. When these parts come together to form a new structured and organized system, there first occurs a deconstruction and then a subsequent restructuring.

As we have explored philosophy and science's long journey toward the understanding of matter and life, we have also indicated the essential importance of closing the causal and explanatory cycles (circular causality) for overcoming the classical and unsatisfactory understanding of causality. Circular causality, which was originally conceptualized in the field of cybernetics, is that which occurs and that generally characterizes organized systems. As Morin demonstrated, and as we wish to stress, this circular causality is not exclusively a characteristic of living system but is, rather, a fundamental characteristic of systemics. Systemic chemical explanations employ circular causality in the inanimate, organic, and biochemical worlds and, for this reason, chemical explanations as they have been conceptualized over the centuries represent systemic explanations *ante litteram*.[6]

We have chosen to focus on these distinct approaches to structuring or organizing the world for two reasons. On the one hand, our choice relates to the general philosophical problem that concerns entities and processes (or transformations) in science. On the other hand, we believe that the scientific disciplines, particularly, physics, chemistry, and biology, employ different conceptions of entity and process that, in turn, inform the different types of explanations utilized by each of these disciplines.

Entities and Processes in Science and Their Use in Different Disciplines

The concepts of entity and process (or transformation) have always functioned as the binomial explanation for the persistence and variability of the world around us, from the inanimate to the animate, from human beings to their productions.

---

[4] El-Hani, "On the Reality of Emergents," 58.
[5] Dupré, "It Is Not Possible to Reduce Biological Explanations to Explanations in Chemistry and/or Physics," 32.
[6] Villani, "La chimica: una scienza della complessità sistemica *ante litteram*."

In philosophy, this combination is usually represented by the concepts of being and becoming, and the problem of explaining the presence of both stability and change was at the core of Western philosophy from its inception. Here, we must first emphasize that the distinction between entity, which represents the stability of being, and transformation, which represents the change at the core of becoming, is not absolute from either a philosophical or scientific perspective.[7]

An entity is defined as a set of properties presented to our instruments or our senses as being static. However, such properties and thus the whole entity are, in fact, slowly transforming through the process of detection by the measuring instruments used to observe them. Philosophically and scientifically, nothing in the universe is truly static, and things do not all change at the same speed. Those objects that are conceived as entities are precisely those objects whose transformations occur at a relatively slow speed. The different transformation speeds determine the time scale that, along with the scales of dimension and energy, permit us to simplify the complex that is evolving through static aspects and dynamic transformations. It is this time scale that also permits the conceptual differentiation of the various processes at play, identifying the slow processes and obtaining properties and, thus, entities that are static relative to those properties exhibiting faster transformations. This is true for living beings, for inanimate matter, and for concepts as well.

In physics, an entity is described by an n-tuple of observable physical quantities, and, as discussed in Chapter 10, these are divided into two groups: those that define the entity and those that characterize the state. The latter quantities are called generalized coordinates of an appropriate space, which is called the configuration space. This space is essentially static, and, therefore, it can only be used to describe the equilibrium states of the system. If we also define the generalized velocities—that is, the derivatives with respect to the time of the generalized coordinates—we can define a phase space in which each point identifies the dynamic state of the system.

In chemistry, the approach is different. A chemical entity is a determined individual with characteristics of its own, and it is so particular that it is given a proper name. Such an entity can be identified at both the macroscopic level, in the realm of pure chemical elements and compounds, and the microscopic level, in the atomic and molecular realms. For example, acetylsalicyclic acid is a distinct and unique chemical compound, with defined chemical properties and well-characterized structured molecules. Although there are no physical substances, there are millions of chemical individuals that explain the different characteristics of the material world through their distinctive properties.

---

[7] Villani, "Ruolo degli enti e dei processi in Chimica e in Fisica."

Chemistry provides an image of the material world as a tumultuous, highly complex world that in no way resembles the relatively simple world of physics, which is made of material points and nameless objects without friction. The chemical world more closely resembles the biological world, with its species, classifications, and problems of nomenclature. Chemistry therefore functions as the true link between the inanimate and animate worlds, allowing a "soft" passage from the simple physical realm to the complex biological realm.

It is not simply the concept of chemical entity that is distinct from entities in physics but also the concept of chemical process. The concept of process is of considerable importance in chemistry, which has actually coined an ad hoc term for describing processes—chemical reaction. In physics, it makes no sense to talk about molecules or even to consider atoms in chemical reactions. Within the limits of quantum mechanics, the reactive atomic/molecular system is considered as a set of atomic nuclei and electrons with certain positions. For chemists, however, all systems consist of atoms and molecules and, in the macroscopic realm, they consist of chemicals, and this also applies to reactive systems.

In physics, a reaction is a *transformation* process that modifies the average positions of atomic nuclei and, to an even greater extent, the distribution of electrons. This process usually takes place by inserting energy or matter into the system, and it occurs over time, with its own energy profile. In practice, physicists regard chemical reactions as perturbations that transform the overall equilibrium condition of the system, leading from the initial to the final equilibrium. According to this perspective, there is no change in the system but only a different equilibrium state of that system. If the system is open and there is either input or output of matter, some formal complication arises. But from the conceptual point of view, there is no change.

From a chemical perspective, however, a reaction always implies entities such as molecules or chemicals that are destroyed and others that are generated or, when considering atoms and elements, entities that are disorganized and reorganized. Obviously, the atoms that are present in the reagents are not materially "destroyed," since atoms are *invariant* for reactivity. However, the atoms have become part of new entities; that is, they have become part of the products of the reaction. This means that the initial and final systems are made up of different entities and are, therefore, two different systems. Following a chemical reaction means identifying the species that react, the intermediate species of reactions if there are any, and those that are finally produced. A chemical reaction is, in fact, a transition between two different systems rather than a transformation of a single system. The concept of the reaction mechanism also changes in the chemical field through the search for new entities, reaction intermediates or activated complexes, that have the fundamental property of being unstable molecular systems with intermediate characteristics between reagents and products.

The conceptual transformation or transition pair synthesizes the difference between the two approaches, physical and chemical, with respect to the conceptual problem of change in general and of chemical reaction in particular. In this context, we can also examine biology, which uses the concept of species and, after Darwin, also faces the problem of the transformability of species. We can note in passing that biology uses the term *evolution* in order to highlight that the species are not fixed in time even if, perhaps, it would be more appropriate to highlight "transition" between species as entities, discontinuity rather than continuity. In summary, on a hypothetical scale from physics through chemistry to biology, we can state that the importance of the concept of entity increases from physics to biology and that the concept of transformation moves in the opposite direction. This observation allows us to rationalize at least two things: the importance of mathematics in all three disciplines and the types of explanations they use. A discussion of both would take us on a tangent from our main focus here, but interested readers are referred to Villani's *Complesso e organizzato. Sistemi strutturati in fisica, chimica, biologia ed oltre*.[8]

We take the position in this book that chemistry's use of the concepts of entity and process is more balanced than the functional approach that has been proposed for contemporary biology. This is even more so in the case of the systemic chemical perspective we are proposing. We must, therefore, turn to the systemic and nonreductionist chemical perspective to achieve the explanatory richness of both the concept of entity and the concept of process in living and nonliving systems.

Another aspect that must be considered is the intrinsically dynamic nature of systems. As already mentioned in the Introduction, the system has a network of internal dynamic interactions, as well as a network of dynamic interactions with external entities that are present in the environment and with the environment in general. This is an essential feature of a dynamic systemic chemical approach to living organisms such as our own, and it must be taken into account when comparing our approach with that of functional biology. To clarify this point, let us consider Dupré and Nicholson's claim that "[i]f physics directs us towards process metaphysics, and if there are additional reasons for thinking that chemistry is, likewise, amenable to a process ontological interpretation, it would be surprising to find that biology pushes us in the opposite direction, towards an ontology of substances."[9]

---

[8] Villani, *Complesso e organizzato. Sistemi strutturati in fisica, chimica, biologia ed oltre*.

[9] Nicholson and Dupré, "Introduction," in *Everything Flows. Towards a Processual Philosophy of Biology*, 15.

Focusing on chemistry, Guttinger studies the case of macromolecules and, in particular, those that are intrinsically disordered (such as those discussed in Chapter 11). He arrives at the conclusion that

> [t]he structure that the protein (or any molecule, for that matter) adopts is therefore the outcome of a complex process, which takes place within a larger dynamic system. Within this system it is not clearly defined what should be seen as "internal" and what as "external", since the boundaries between the entity of interest and its environment are blurred. The capacity to adopt a particular fold is therefore not something that the protein simply possesses and that is then triggered or activated by some external input from the environment but it is, like the catalytic power Stein discusses, or like the capacities of a termite colony, an integrated capacity that emerges from within an integrated whole.[10]

Guttinger's analysis repeats our own discussion in Chapter 11, where we examined at length intrinsically disordered proteins and the folding of proteins, since this process illustrates that a macromolecule is a system that behaves as a whole. It is evident that all the components act together and dynamically in the formation of a new system. However, after the new system is formed, we must consider its global action within the environment. Once we evaluate the level of complexity of the macromolecule, we find that its properties are global but not necessarily static. As we have pointed out several times, the system is an intrinsically dynamic entity, and, at the component level, there are properties that cannot be traced back to the individual parts.

The fact that, in the experiment discussed, a number of water molecules must be added to describe the environment and its interaction with the macromolecule does not alter the terms of systemic interaction. The description must always include the internal and specific characteristics of the macromolecule, those of the water molecule, and the interaction of these structured entities. As stressed in Chapter 11, the complexification of the proteome requires that we understand "more chemistry," not less chemistry, and such a chemistry must be one that uses a systemic approach. Furthermore, when examining cells, sets of reactions (cyclical or not) should be foregrounded over individual reactions, since this enriches the dynamic systems problems that our approach seeks to resolve.

---

[10] Guttinger, "A process ontology for macromolecular Biology," 318.

Chemical Explanation as a Model for Other Disciplines

The systemic chemical approach analyzed and advocated here has some peculiarities but also some general characteristics that are common to all systemic approaches. With regard to the general characteristics, the detailed treatment that these have received in this book should suffice and not require further expansion in this conclusion. However, although the peculiarities have also been extensively discussed, there is one aspect of these that we wish to emphasize here. This aspect is linked to the way in which chemistry uses the conceptual terms entity and process. As we have already stated, chemistry is that scientific discipline that employs these two essential concepts in a more balanced fashion, and this is our principal reason for proposing chemistry as a model for the humanities and social sciences.

In this book, we have discussed both theory and practice, and we wish to repeat that, unlike physics, chemistry does not seek explanations that subsume natural entities and processes under universal laws. Instead, chemical explanations seek to understand entities and processes through the static and dynamic properties of constituents (molecules at the microscopic level and compounds at the macroscopic level) within a specific context and how such properties determine properties and behavior at higher levels of material reality. In chemistry, there is a moment of analysis during which the components are identified, followed by a moment of synthesis during which the components are brought together to create a new substance. This process reflects chemical practice at all levels, from research to industry.

At a microscopic level, the concept of molecule is both a scientific and a philosophical resource for chemistry. Although we will not dwell further here on the scientific importance of the molecule/compound binomial for chemical transformations, we stress that this combination is present in almost all chemical practice. We will, however, say a few words regarding the philosophical importance of molecule because this has been an undervalued aspect of this concept, even in chemistry itself.

Why is the molecular realm so significant that it transcends being merely a pure scientific concept? We argue that the concepts of molecule and macromolecule, as structured and dynamic entities, are philosophical notions as well as scientific ones, and the level of complexity of molecules presents some peculiar problems. In a list of the levels of complexity of material reality, that which leads from elementary particles to the macroscopic level, we see that the molecular level is located immediately before the bifurcation between the inanimate and animate realm.

The study of molecules is the level immediately preceding the study of living systems, and so, as is highlighted in biochemistry, understanding molecules is fundamental to understanding living matter. However, the molecular level also immediately precedes the level of macroscopic inanimate material objects, so that molecules are also a reference point in the study of inanimate macroscopic objects. As an example, we see the role played by geochemistry in the study of minerals, rocks, and the like. Thus, an essential philosophical characteristic of chemistry is that it represents a unique and specific language enabling the study of material reality, both inanimate and animate. It is, therefore, essential to emphasize that no modern scientific discipline can dispense with chemistry.

Another peculiarity of chemistry is that it studies the material world in all of its qualitative richness, a world in which millions of entities are so distinct from one another that each is given an individual name. It is this characteristic that makes chemistry particularly useful for explaining both the complex macroscopic world in general and, more specifically, the even more complex world of living matter. The breadth of chemistry's epistemic reach is a direct consequence of the structuring that occurs at the molecular level. Thus, the attention that we have placed on the importance of the concept of molecule, for chemistry specifically and for science more generally, has been justified.

Let us now examine how the systemic chemical approach that we advocate can serve as a model for the humanities and the social sciences. The systemic approach has long been present in the field of sociology. As Jiří Šubrt states, "the main task of a systemic approach in sociology should be to analyse systemic processes at the macro-social level . . . Applying functionalist methodology thus involves examining the individual parts of a system (subsystems) with regard to their specific contributions (i.e. functions) to maintaining the whole, placing emphasis on integrity and equilibrium."[11] Šubrt then states that the optimism and enthusiasm that initially accompanied the systemic approach in sociology have considerably waned. He also notes that systemic approaches still survive and continue to develop in the areas of the theory of networks, entropy, chaos, and synergy concepts, which address the issues of self-organization or creation and of stability and disappearance of organized and spontaneously emerging temporal and spatial structures.

Sociology, like the other social sciences and the humanities, continues to focus on the procedural or physical approaches to identifying the specific characteristics of sociological entities. In practice, however, sociology also focuses on the network of identical nodes that interact according to different patterns. This method is different from the chemical approach that we advocate, since even a few nodes with specific characteristics can generate complex systems that

---

[11] Šubrt, "Systemic Approach in Sociology: Reflections on Its Development, Current Status and Possibilities," 81–82

display their own unique characteristic and properties. The systemic model of chemical explanation can contribute to the social sciences and the humanities by demonstrating how an approach in which entities and processes are present on an equal footing can help us understand our complex living world more fully and deeply.

Bibliography

Abbas, Abul., Andrew Lichtman, and Shiv Pillai. *Cellular and Molecular Immunology.* Philadelphia: Elsevier, 2012.

Agutter, Paul S., and Denys N. Wheatley. *About Life: Concept in Modern Biology.* Dordrecht: Springer, 2007.

Amaldi, Francesco. "Evoluzione della selezione." ' In *Selezione e selezionismi*, edited by Saverio Forestiero and Massimo Stanzione, 50–60. Milan: Franco Angeli, 2011.

Anderson, Philip W. "More Is Different." *Science* 177 (1972): 393–396.

Anfinsen, Christian Boehmer. "Principles That Govern the Folding of Protein Chains." *Science* 181 (1973): 223–230.

Anstey, Peter. "Essences and Kinds." In *Oxford Handbook of Philosophy in Early Modern Europe*, edited by Catherine Wilson and Desmond M. Clarke, 11–31. Oxford: Oxford University Press, 2011.

Atlan, Henri. "Sul rumore come principio di auto-organizzazione." In *Teorie dell'evento*, edited by Edgar Morin, 35–58. Milan: Bompiani, 1972.

Baedke, Jan. *Above the Gene, Beyond Biology: Toward a Philosophy of Epigenetics.* Pittsburgh: University of Pittsburgh Press, 2018.

Balzani, Vincenzo, Margherita Venturi, and Alberto Credi. *Molecular Devices and Machines. A Journey into the Nanoworld.* Weinheim: Wiley-VCH, 2003.

Banchetti-Robino, Marina Paola. "Ontological Tensions in Sixteenth and Seventeenth Century Chemistry: Between Mechanism and Vitalism." *Foundations of Chemistry* 13, No. 3 (2011): 173–186.

Banchetti-Robino, Marina Paola. "The Ontological Function of First-Order and Second-Order Corpuscles in the Chemical Philosophy of Robert Boyle: The Redintegration of Potassium Nitrate." *Foundations of Chemistry* 14, No. 3 (2012): 221–234.

Banchetti-Robino, Marina Paola. "From Corpuscles to Elements: Chemical Ontologies from Van Helmont to Lavoisier." In *Philosophy of Chemistry: Growth of a New Discipline*, edited by Lee McIntyre and Eric Scerri, 141–154. Dordrecht: Springer, 2014.

Banchetti-Robino, Marina Paola. "The Relevance of Boyle's Chemical Philosophy for Contemporary Philosophy of Chemistry." In *The Philosophy of Chemistry: Practices, Methodologies, and Concepts*, edited by Jean-Pierre Llored, 240–265. Dordrecht: Springer, 2014.

Banchetti-Robino, Marina Paola. "Van Helmont's Hybrid Ontology and Its Influence on the Chemical Interpretation of Spirit and Ferment." *Foundations of Chemistry* 18 (2015): 103–112.

Banchetti-Robino, Marina Paola, and Jean-Pierre Llored. "Reality without Reification: Philosophy of Chemistry's Contribution to Philosophy of Mind." In *Oxford Handbook on the Philosophy of Chemistry*, edited by Grant Fischer and Eric Scerri, 83–110. Oxford: Oxford University Press, 2016.

Banchetti-Robino, Marina Paola. *The Chemical Philosophy of Robert Boyle: Mechanism, Chymical Atoms, and Emergence.* Oxford: Oxford University Press, 2020.

Bateson, Gregory. "Cybernetic Explanation." *American Experimental Scientist* 10, No. 8 (1967): 29–32.

Bechtel, William. *Mental Mechanisms: Philosophical Perspectives on Cognitive Neuroscience*. London: Routledge, 2008.

Bellone, Enrico. *I nomi del tempo: La seconda rivoluzione scientifica e il mito della freccia temporale*. Turin: Bollati Boringhieri, 1999.

Bensaude-Vincent, Bernadette, and Isabelle Stengers. *Histoire de la* Chimie. Paris: La Découverte, 1992.

Bensaude-Vincent, Bernadette, and Isabelle Stengers. *A History of Chemistry*. Cambridge, MA: Harvard University Press, 1996.

Bergman, Tobern Olaf. *A Dissertation on Elective Attractions*. London: J. Murray, 1785.

Bergman, Tobern Olaf. *Traité de Affinités Chymiques ou Attractions Éléctives*, translated by François Joseph Bonjour. Paris: Buisson, 1788.

Bertalanffy, Ludwig von. "Der Organismus als physikalisches System betrachtet." *Natutrwissenschaften* 28 (1940): 521–531.

Bertalanffy, Ludwig von. *Theoretische Biologie*. Berlin: Gebruder Borntraeger, 1940.

Bertalanffy, Ludwig von. *Das Biologische Weltbild*. Bern: A. Franche, 1949.

Bertalanffy, Ludwig von. "Zu einer allgemeinen Systemlehre." *Biologia Generalis* 195 (1949): 114–129.

Bertalanffy, Ludwig von. "The Theory of Open Systems in Physics and Biology." *Science* 111 (1950): 23–29.

Bertalanffy, Ludwig von. *Problems of Life*. New York: Harper, 1952.

Bertalanffy, Ludwig von. *General System Theory*. New York: George Braziller, 1968.

Berthollet, Claude-Louis. *Séances des écoles normales recueillies par des sténographes et revues par les professeurs*. Paris: Écoles Normales Supérieures, 1794.

Berthollet, Claude-Louis. *Recherches sur les lois de l'affinité*. Paris: Baudouin, 1801.

Berthollet, Claude-Louis. *Éssai de statique* chimique. Paris: Demonville, 1803.

Berthelot, Marcellin. *Chimie Organique Fondé sur la Synthèse*. Paris: Mallet–Bachelier, 1860.

Berthelot, Marcellin. "Sur la fermentation glucosique du sucre de canne." *Comptes rendus de l'Académie des Sciences* 50 (1860): 980–984.

Boas Hall, Marie. "The History of the Concept of Element." In *John Dalton and the Progress of Science*, edited by D. S. L. Cardwell, 21–39. Manchester: Manchester University Press, 1968.

Born, Max. "Zur Quantenmechanik der Stossvorgänge." *Zeitschrift für Physik* 37 (1926): 863–867.

Born, Max. "Zur Quantenmechanik der Stossvorgänge." *Zeitschrift für Physik* 38 (1926): 803–827.

Born, Max. "Zur Wellnmechanik der Stossvorgänge." *Göttinger Nachrichten* (1926): 146–160.

Brigandt, Ing, and Alan Love. "Reductionism in Biology." In *The Stanford Encyclopedia of Philosophy*, edited by Edward N. Zalta. Stanford, CA: Stanford University Press, 2012. Available at: https://plato.stanford.edu/archives/sum2012/entries/reduction-biology.

Buchner, Eduard. *Cell-Free Fermentation: Nobel Lecture* (December 11, 1901). Available at: http://www.nobelprize.org/uploads/2018/06/buchner-lecture.pdf.

Buiatti, Marcello. *Lo stato vivente della materia. Le frontiere della nuova biologia*. Turin: UTET, 2000.

Butlerov, Alexander M. *Works*. Moscow: 1953.

Bykov, G. V. "Origin of the Theory of Chemical Structure." *Journal of Chemical Education* 39 (1962): 220–224.

Calissano, Pietro. *Neuroni. Mente ed evoluzione.* Milan: Garzanti, 1992.

Canguilhem, Georges. "La conoscenza della vita." In *L'epistemologia francese contemporanea: Per un razionalismo aperto,* edited by Carlo Vinti, 239–242. Rome: Città Nuova Editrice, 1977.

Carvalho Ramos, Maurício de. "Organic Monadology in Maupertuis." *Advances in Historical Studies* 4 (2015): 17–28.

Cassirer, Ernst. *The Philosophy of the Enlightenment.* Translated by Fritz C. A. Köln and James P. Pettegrove. Princeton, NJ: Princeton University Press, 1951.

Cerruti, Luigi. *Bella e potente. La chimica del Novecento fra scienza e società.* Rome: Editori Riuniti, 2003.

Chisholm, Roderick M. "The Contrary-to-Fact Conditional." *Mind* 55 (1946): 289–307.

Christie, John R., and Maureen Christie. "Chemical Laws and Theories: A Response to Vihalemm." *Foundations of Chemistry* 5 (2003): 165–174.

Christie, Maureen, and John R. Christie. "'Laws' and 'Theories' in Chemistry Do not Obey the Rules." In *Of Minds and Molecules: New Philosophical Perspectives on Chemistry,* edited by Nalini Bhushan and Stuart Rosenfeld, 34–50. Oxford: Oxford University Press, 2000.

Clericuzio, Antonio. *Elements, Principles, and Corpuscles: A Study of Atomism and Chemistry in the Seventeenth Century.* Dordrecht: Kluwer Academic Publishers, 2000.

Clericuzio, Antonio. "The Internal Laboratory: The Chemical Reinterpretation of Medical Spirits in England (1650–1680)." In *Alchemy and Chemistry in the 16th and 17th Centuries,* edited by Piyo Rattansi, 51–83. Dordrecht: Kluwer Academic Publishers, 1994.

Cohen, Paul S., and Stephen M. Cohen. "Wöhler's Synthesis of Urea: How Do the Textbooks Report It?" *Journal of Chemical Education* 73 (1996): 883–886.

Constable, Edwin C. and Catherine E. Housecroft. "Chemical Bonding: The Journey from Miniature Hooks to Density Functional Theory." *Molecules* 25, No. 11 (2020): 1–44.

Corliss, John O. "The Protozoon and the Cell: A Brief Twentieth Century Overview." *Journal of the History of Biology* 22 (1989): 307–323.

Craver, C. F. "Beyond Reduction: Mechanisms, Multifield Integration and the Unity of Neuroscience." *Studies in History and Philosophy of Biological and Biomedical Sciences* 36 (2005): 373–395.

Dalton, John. "On the Absorption of Gases by Water and Other Liquids." In *Memoirs of the Literary and Philosophical Society of Manchester,* Second Series, 1 (1805): 271–287.

Dalton, John. *A New System of Chemical Philosophy.* London: R. Bickerstaff, 1808.

Davy, Humphrey. "Some Experiments and Observations on a New Substance Which Becomes a Violet-Coloured Gas by Heat." *Philosophical Transactions of the Royal Society* 104, No. 104 (1814): 74–93.

Dawkins, Richard. *The Blind Watchmaker: Why the Evidence of Evolution Reveals a Universe without Design.* New York: W. W. Norton, 1986.

Darwin, Charles. *The Life and Letters of Charles Darwin: Including Autobiographical Chapter.* Edited by Francis Darwin. New York: Appleton, 1896.

Debus, Alan G. *The Chemical Philosophy.* New York: Dover Publications, 1977.

Descartes, René. *Principia philosophiae.* Frankfurt-am-Main: Friderici Knochii & Filii, 1722, 51.

Descartes, René. *Oeuvres Philosophiques de Descartes.* Edited by M. L. Aimé-Martin. Paris: Auguste Desrez, 1838.

Descartes, René. *Oeuvres de Descartes.* Edited by Charles Adam and Paul Tannery. Paris: Léopold Cerf, 1905.

Descartes, René. *The Correspondence between Descartes and Henricus Regius / De briefwisseling tussen Descartes en Henricus Regius.* Edited by Jan Jacobus Frederik Maria Bos. *Quaestiones Infinitae: Publications of the Department of Philosophy,* Volume 37. Utrecht: Utrecht University, 1969.

Dirac, Paul. "Quantum Mechanics of Many-Electron Systems." *Proceedings of the Royal Society of London* A 123 (1929): 714–733.

Dray, W. H. *Laws and Explanation in History.* Oxford: Oxford University Press, 1957.

Duhem, Pierre. *"Mixture and Chemical Combination" and Related Essays.* Edited, translated, and with an introduction by Paul Needham. Dordrecht: Springer, 2002.

Dupré, John. "It Is Not Possible to Reduce Biological Explanations to Explanations in Chemistry and/or Physics." In *Contemporary Debates in Philosophy of Biology,* edited by Francisco J. Ayala and Robert Arp, 32–47. Singapore: Wiley-Blackwell, 2010.

Egdell Russell G., and Elizabeth Bruton. "Henry Moseley, X-ray Spectroscopy and the Periodic Table." *Philosophical Transactions of the Royal Society* A 378, No. 2180 (2020). Available at: https://royalsocietypublishing.org/doi/10.1098/rsta.2019.0302.

El-Hani, Charbel N. "On the Reality of Emergents." *Principia* 6 (2002): 51–87.

Feingold, Mordekai. "Mathematicians and Naturalists, Sir Isaac Newton and the Royal Society." In *Isaac Newton's Natural Philosophy,* edited by Jed Z. Buchwald and I. Bernard Cohen, 77–102. Cambridge, MA: MIT Press, 2001.

Foerster, Heinz von. "On Self-Organizing Systems and Their Environments." In *Self-Organizing Systems: Proceedings of an Interdisciplinary Conference,* edited by Marshall C. Yovits, and Scott Cameron, 1–375. New York: Pergamon Press, 1960.

Franklin, Rosalind E., and R. G. Gosling. "Molecular Configuration in Sodium Thymonucleate." *Nature* 171 (1953): 740–741.

Fujii, Kiyohisa. "The Berthollet-Proust Controversy and Dalton's Chemical Atomic Theory 1800–1820." *The British Journal for the History of Science* 19, No. 2 (1986): 177–200.

Gagliasso, Elena. *Verso un'epistemologia del mondo vivente. Evoluzione e biodiversità tra legge e narrazione.* Milan: Guerini, 2001.

Garber, Daniel. "Physics and Foundations." In *The Cambridge History of Science,* edited by Katharine Park and Lorraine Daston, 21–69. Cambridge: Cambridge University Press, 2006.

Geison, Gerald L. "Pasteur on Vital versus Chemical Ferments: A Previously Unpublished Paper on the Inversion of Sugar." *Isis* 72 (1981): 425–455.

Ghibaudi, Elena, Luigi Cerruti, and Giovanni Villani. "Structure, Shape, Topology: Entangled Concepts in Molecular Chemistry." *Foundations of Chemistry* 22 (2020): 279–307.

Gilbert, Scott F., Jan Sapp, and Alfred I. Tauber. "A Symbiotic View of Life: We Have Never Been Individuals." *The Quarterly Review of Biology* 87, No. 4 (2012): 325–341.

Gilson, Étienne. *Biofilosofia Da Aristotele a Darwin e ritorno.* Genoa-Milan: Marietti, 2003.

Goodman, Nelson. "The Problem of Counterfactual Conditionals." *The Journal of Philosophy* 44 (1947): 113–128.

Graeme, K. Hunter. *Vital Forces: The Discovery of the Molecular Basis of Life.* New York: Academic Press, 2000.

Greco, Pietro. *Evoluzioni. Dal Big Bang a Wall Street. La Sintesi Impossibile.* Naples: CUEN, 1999.

Gregory, Joshua C. "The Animate and Mechanical Models of Reality." *Journal of Philosophical Studies* 2, No. 7 (1927): 301–314.

Guttinger, Stephan. "A Process Ontology for Macromolecular Biology." In *Everything Flows. Towards a Processual Philosophy of Biology*, edited by Daniel J. Nicholson and John Dupré, 303–320. Oxford: Oxford University Press, 2018.

Halbwachs, Francis. "L'histoire de l'explication en physique." *Colloque de l'académie internationale de la philosophie des sciences avec le concours du Centre international d'épistémologie génétique (Genève, 25–29 September, 1970)*, edited by Leo Apostel, 72–102. Paris: Flammarion, 1973.

Hegel, Georg Wilhelm Friedrich. *Encyclopedia of the Philosophical Sciences, Part I: The Logic.* Translated and edited by Klaus Brinkmann and Daniel O. Dahlstrom. Cambridge: Cambridge University Press, 2020.

Hein, George E. "Kekulé and the Architecture of Molecules." In *Kekulé Centennial*, edited by O.Theodor Benfey, 1–12. Washington, DC: American Chemical Society, 1966.

Hempel, Carl G. "The Function of General Laws in History." *The Journal of Philosophy* 39 (1942): 35–48.

Hempel, Carl G. "The Logic of Functional Analysis." In *Aspects of Scientific Explanation and Other Essays*, 297–330. New York: The Free Press, 1970.

Hempel, Carl G. "Deductive-Nomological versus Statistical Explanation." In *Minnesota Studies in the Philosophy of Science: Volume 3*, edited by Herbert Feigl and Grover Maxwell, 98–169. Minneapolis: University of Minnesota Press, 1962.

Hempel, Carl G. *Aspects of Scientific Explanation and Other Essays in the Philosophy of Science.* New York: The Free Press, 1965.

Hempel, Carl G. "Explanation in Science and in History." In *Frontiers in Science and Philosophy*, edited by Robert G. Colony, 276–284. Pittsburgh, PA: University of Pittsburgh Press, 1963.

Hempel, Carl G. *Philosophy of Natural Science.* Englewood Cliffs, NJ: Prentice Hall, 1966.

Hempel, Carl G. *Filosofia delle scienze naturali.* Bologna: il Mulino, 1972.

Hempel, Carl G., and Paul Oppenheim. "Studies in the Logic of Explanation." *Philosophy of Science* 15, No. 2 (1948): 135–175.

Hendry, Robin Findlay. "Lavoisier and Mendeleev on the Elements." *Foundations of Chemistry* 7 (2005): 31–48.

Hendry, Robin Findlay. "Elements, Compounds, and Other Chemical Kinds." *Philosophy of Science* 73 (2006): 864–875.

Hendry, Robin Findlay. "Antoine Lavoisier (1743–1794)." In *Philosophy of Chemistry*, edited by Andrea I Woody, Robin Findlay Hendry, and Paul Needham, 63–70. The Netherlands: Elsevier, 2012.

Holmes, Frederic L. "From Elective Affinities to Chemical Equilibria: Berthollet's Law of Mass Action." *Chymia* 8 (1962): 105–145.

Hunter, Graeme K. *Vital forces: The Discovery of the Molecular Basis of Life.* New York: Academic Press, 2000.

Jablonka, Eva, and Marion J. Lamb. *Epigenetic Inheritance and Evolution.* Oxford: Oxford University Press, 1995.

Jablonka, Eva, and Marion J. Lamb. *Evolution in Four Dimensions. Genetic, Epigenetic, Behavioral, and Symbolic Variation in the History of Life.* Cambridge, MA: MIT Press, 2005.

Jenuwein, Thomas, and C. David Allis. "Translating the Histone Code." *Science* 293 (2001): 1074–1079.

Joly, Bernard. "Le cartésianisme de Boyle du point de vue de la chimie." In *La Philosophie Naturelle de Robert Boyle*, edited by Myriam Dennehy and Charles Ramond, 139–155. Paris: Vrin, 2009.

Kant, Immanuel. *Metaphysical Foundations of Natural* Science, translated by Michael Friedman. Cambridge: Cambridge University Press, 2004.

Kapoor, Satish C. "Berthollet, Proust, and Proportions." *Chymia* 10 (1965): 53–110.

Kargon, Robert Hugh. *Atomism in England from Hariot to Newton*. Oxford: Oxford University Press, 1966.

Kauffman, Stuart. *At Home in the Universe: The Search for Laws of Self-Organization and Complexity*. Oxford: Oxford University Press, 1995.

Kekulé, August. "Ueber die s.g. gepaarten Verbindungen und die Theorie der mehratomigen Radicale." *Annalen der Chemie und Pharmacie* 104, No. 2 (1857): 129–150.

Kekulé, August. "Sur l'atomicité des éléments." *Comptes Rendus Hebdomadaire de l'Académie des Sciences* 58 (1864): 510–514.

Kekulé, August. "Sur la constitution des substances aromatiques." *Bulletin de la Société Chimique de France* 3 (1865): 98–110.

Knight, David. *Atoms and Elements: A Study of Theories of Matter in England in the Sixteenth Century*. London: Hutchinson, 1967.

Knight, David. "John Dalton (1766–1844)." In *Philosophy of Chemistry*, edited by Andrea I Woody, Robin Findlay Hendry, and Paul Needham, Handbook of the Philosophy of Science Series, 71–78. The Netherlands: Elsevier, 2012.

Kossel, Albrecht. "The Chemical Composition of the Cell." *Harvey Lecture Series* 7 (1911): 33–51.

Krebs, Hans. *The Citric Acid Cycle: Nobel Lecture* (December 11, 1953). Available at: https://www.nobelprize.org/prizes/medicine/1953/krebs/facts/

Kull, Kalevi. "Vegetative, Animal, and Cultural Semiosis: The Semiotic Threshold Zones." *Cognitive Semiotics* 4 (2009): 8–27.

Kutter, Elizabeth. "Analysis of Bacteriophage T4 Based on the Completed DNA Sequence." In *Integrative Approaches to Molecular Biology*, edited by Julio Collado-Vides, Boris Magasanik, and Temple F. Smith, 13–28. Cambridge, MA: MIT Press, 1996.

La Mettrie, Julien Offray de. *L'Homme Machine*, edited by A. Vartanian. Princeton, NJ: Princeton University Press, 1960.

Laplace, Pierre-Simon. "*Essai philosophique sur les probabilités.*"In *Oeuvres complètes de Laplace*, Vol. VII, Paris: Gauthier-Villars, 1878.

Laplace, Pierre-Simon. *Oeuvres complètes de Laplace*. Paris: Gauthier-Villars, 1878–1912.

Lavoisier, Antoine-Laurent. *Traité Élémentaire de Chimie*. Paris: Cuchet Libraire, 1789.

Lavoisier, Antoine. *Traité Élémentaire de Chimie*. Italian translation by Vincenzo Dandolo. Venice: Antonio Zatta e Figli,1802.

Lederman, Leon, and Dick Teresi. *The God Particle: If the Universe Is the Answer, What Is the Question?* New York: Dell Publishing, 1993.

LeGrand, Homer Eugene. *Berthollet and the Oxygen Theory of Acidity*. Madison: University of Wisconsin Press, 1971.

Lehman, Christine. "Mid-Eighteenth Century Chemistry in France as Seen through Student Notes from the Course of Gabriel-François Venel and Guillaume-François Rouelle." *Ambix* 56 (2009): 163–189.

Lehn, Jean-Marie. *Supramolecular Chemistry. Concepts and Perspectives*. Weinheim: Wiley-VCH, 1995.

Leibniz, Gottfried Wilhelm. *Leibnizens mathematische Schriften*. Edited by Carl Immanuel Gerhardt. Berlin: Eidmann, 1850.

Leibniz, Gottfried Wilhelm. *The Early Mathematical Manuscripts of Leibniz*, translated from the Latin by Carl Immanuel Gerhardt and edited by J. M. Child. London: The Open Court Publishing Co., 1920.

Leibniz, Gottfried Wilhelm. *Discourse on Metaphysics* and *The Monadology*, translated by George R. Montgomery. New York: Dover Publications, 2005.

Leicester, Henry M. *The Historical Background of Chemistry*. New York: Dover Publications, 1956.

Levere, Trevor H. *Transforming Matter: A History of Chemistry from Alchemy to the Buckyball*. Baltimore, MD: Johns Hopkins University Press, 2001.

Lewis, David. "Introduction to an English Translation of 'On the Different Explanations of Certain Cases of Isomerism' by Aleksandr Butlerov." *Bulletin for the History of Chemistry*, 40 (2015): 9–12.

Liebig, Justus, and Joseph L. Gay-Lussac. "Analyse du fulminate d'argent." *Annales de Chimie et de Physique* 25 (1824): 285–311.

Madan, H. G. "Remarks on Some Points in the Nomenclature of Salts." *Journal of the Chemical Society* 23 (1870): 22–28.

Maupertuis, Pierre Louis. *Système de la nature: Essai sur la formation des corps organisés*. In *Œuvres*. Lyon, 1756.

Maupertuis, Pierre Louis. *Œuvres*. Hildesheim: Geog Olms, 1965.

Mayr, Ernst. *Storia del pensiero biologico. Diversità, evoluzione, eredità*. Turin: Bollati Boringhieri, 1990.

McLaughlin, Brian P. "The Rise and Fall of British Emergentism." In *Emergence: Contemporary Reading in Philosophy and Science*, edited by Mark A. Bedau, and Paul Humphreys, 19–60. Cambridge: MIT Press, 2008.

Meinel, Christoph. "Early Seventeenth-Century Atomism: Theory, Epistemology and the Insufficiency of Experiment." *Isis* 79 (1988): 68–103.

Meinel, Christoph. "Empirical Support for the Corpuscular Theory in the Seventeenth Century." In *Theory and Experiment*, edited by D. Batens and J. P. van Bendegem, 77–92. Dordrecht: D. Reidel, 1988.

Mendeleev, Dmitrii. "The Periodic Law of the Chemical Elements." *Journal of the Chemical Society* 55 (1889): 634–656.

Meyer, Viktor. "Zur Valenz Und Verbindungsfähigkeit des Kohlenstoffs." *Justus Liebigs Annalen der Chemie* 180 (1876): 192–206.

Mill, John Stuart. "A System of Logic, Ratiocinative And Inductive." In *The Project Gutenberg EBook of A System of Logic, Ratiocinative And Inductive by John Stuart Mill*. Online at Project Gutenberg, 2009. Available at: https://www.gutenberg.org/files/27942/27942-h/27942-h.html.

Minkin, Vladimir Ilsaac. "Current Trends in the Development of A. M. Butlerov's Theory of Chemical Structure." *Russian Chemical Bulletin* 61 (2012): 1265–1290.

Mirsky, Alfred Ezra, and Linus Pauling. "On the Structure of Native, Denatured, and Coagulated Proteins." *Proceedings of the National Academy of Sciences* 22 (1936): 439–447.

Monod, Jacques. *Chance and Necessity. An Essay on the Natural Philosophy of Modern Biology*, translated from the French by Austryn Wainhouse. New York: Alfred A. Knopf, 1971.

Moreno, Alvaro. "A Systemic Approach to the Origin of Biological Organization." In *Systems Biology: Philosophical Foundations*, edited by Fred C. Boogerd, Frank J. Bruggeman, Jan-Hendrik S. Hofmeyr, and Hans V. Westerhoff, 243–268. Amsterdam: Elsevier, 2007.

Morin, Edgar. *Il metodo 1. La natura della natura*. Milan: Raffaello Cortina, 2001.

Mukherjee, Siddhartha. *The Gene. An Intimate History*. New Delhi: Scribner, 2016.

Nash, Leonard K. *The Atomic-Molecular Theory*. Cambridge, MA: Harvard University Press, 1967.

Needham, Paul. "An Aristotelian Theory of Chemical Substance." *Logical Analysis and History of Philosophy* 12 (2009): 149–164.

Newman, William R. *Atoms and Alchemy: Chymistry and the Experimental Origins of the Scientific Revolution*. Chicago: University of Chicago Press, 2006.

Newman, William R. "What Have We Learned from the Recent Historiography of Alchemy?" *Isis* 102, No. 2 (2011): 313–321.

Newton, Isaac. *Opticks: A Treatise of the Reflexions, Refractions, Inflexions and Colours of Light*, with a Foreword by Albert Einstein, an introduction by Sir Edmund Whittaker, and a preface by I. Bernard Cohen. New York: Dover Publications, 2012.

Newton, Isaac. *The Principia: Mathematical Principles of Natural Philosophy*. Translated from the Latin by Andrew Motte. New York: Prometheus Books: 1995.

Newton, Isaac. *Unpublished Scientific Papers of Issac Newton: A Selection from the Portsmouth Collection in the University of Cambridge Library*. Edited by A. Rupert Hall and Marie Boas Hall. Cambridge: Cambridge University Press, 1978.

Nicholson, Daniel J., and John Dupré. "Introduction." In *Everything Flows. Towards a Processual Philosophy of Biology*, edited by D. J. Nicholson and J. Dupré, 3–46. Oxford: Oxford University Press, 2018.

Ohno, Susumo. "So Much 'Junk DNA' in Our Genome." In *Evolution of Genetic Systems*, edited by Harold Hill Smith, 366–370. New York: Gordon and Breach, 1972.

Ostwald, Wilhelm. *Lehrbuch Der Allgemeinen Chemie*. Berlin: Nabu Press, 2010.

Palmer, W. G. *A History of the Concept of Valency to 1930*. Cambridge: Cambridge University Press, 2010.

Paneth, Friedrich A. "The Epistemological Status of the Concept of Element (I)." *British Journal for the Philosophy of Science* 13, No. 49 (1962): 1–14.

Paneth, Friedrich A. "The Epistemological Status of the Concept of Element (II)." *British Journal for the Philosophy of Science* 13, No. 50 (1962): 144–160.

Pasnau, Robert. "Form, Substance, and Mechanism." *The Philosophical Review* 113 (January 2004): 31–88.

Pasteur, Louis. "Mémoire de L. Pasteur sur la fermentation appelée lactique." *Comptes Rendus séance de l'Académie des Sciences* 45 (1857): 913–914.

Pauling, Linus. *The Nature of the Chemical Bond*. Ithaca, NY: Cornell University Press, 1939.

Pennazio, Sergio. "Homeostasis: A History of Biology." *Rivista di biologia* 102 (2009): 253–271.

Pépin, François. *La philosophie expérimentale de Diderot et la chimie*. Paris: Classiques Garnier, 2012.

Pichot, André. *Histoire de la notion de vie*. Paris: Gallimard 1993.

Pichot, André. *Expliquer la vie. De l'âme à la molécule*. Versailles: Quae, 2011.

Pinet, Patrice. "La philosophie de la matière de Galilée à Newton." *Revue d'histoire de la pharmacie* 52, No. 341 (2004): 67–82.

Planck, Max. *Nobel Lecture* (June 2, 1920). Available at: https://www.nobelprize.org/prizes/physics/1918/planck/lecture/

Popper, Karl. *Quantum Theory and the Schism in Physics: From the Postscript to The Logic of Scientific Discovery*. Edited by W. W. Bartley III. London: Routledge, 1982.

Prigogine, Ilya. *Étude thermodynamique des Phenomènes Irréversibles*. Paris: Dunod, 1947.

Prigogine, Ilya. *Introduction to Thermodynamics of Irreversible Processes*. Springfield, MA: Charles C. Thomas, 1955.

Prigogine, Ilya. *La nascita del tempo*. Milan: Bompiani, 1988.

Prigogine, Ilya, and Isabelle Stengers. *La Nouvelle Alliance: Métamorphose de la Science*. Paris: Gallimard, 1980.

Pross, Addy. *What Is Life? How Chemistry Becomes Biology*. Oxford: Oxford University Press, 2012.

Proust, Joseph Louis. "Recherches sur le cuivre." *Annales de Chimie* 32 (1799): 26–54.

Proust, Joseph Louis. "Sur les mines de cobalt, nickel et autres." *Journal de Physique* 63 (1806): 364–377.

Prout, William. "On the Relation between Specific Gravities of Bodies in the Gaseous State and the Weights of Their Atoms." *Thomson's Annals of Philosophy* 6 (1815): 321–330.

Prout, William. "Correction of a Mistake in the Essay on the Relation Between Specific Gravities of Bodies in the Gaseous State and the Weights of Their Atoms." *Thomson's Annals of Philosophy* 7 (1816): 111–113.

Pullman, Bernard. *The Atom in the History of Human Thought*. New York: Oxford University Press, 1998.

Putnam, Hilary, and Paul Oppenheim. "Unity of Science as a Working Hypothesis." In *Concepts, Theories, and the Mind-Body Problem*, edited by Herbert Feigl, Michael Scriven, and Grover Maxwell, 3–36. Minneapolis: University of Minnesota Press, 1958.

Quine, Willard van Orman. *Theories and Things*. Cambridge: Cambridge University Press, 1981.

Ramberg, Peter J. "The Death of Vitalism and the Birth of Organic Chemistry: Wöhler's Urea Synthesis and the Disciplinary Identity of Organic Chemistry." *Ambix* 47 (2000): 170–195.

Ramberg, Peter J. *Chemical Structure, Spatial Arrangement: The Early History of Stereochemistry*. London: Routledge, 2003.

Ramsay, O. Bertrand. "Molecules in Three Dimensions (I)." *Chemistry* 47 (1974): 6–9.

Ritter, William Emerson. *The Unity of the Organism: Or the Organismal Conception of Life*. Boston: Gorham Press, 1919.

Rocke, Alan J. *Origins of the Structural Theory in Organic Chemistry, Volume 2*. Madison: The University of Wisconsin Press, 1975.

Rocke, Alan J. "The Reception of Chemical Atomism in Germany." *Isis* 70 (1979): 519–536.

Rocke, Alan J. "Kekulé, Butlerov, and the Historiography of the Theory of Chemical Structure." *British Journal for the History of Science* 14 (1981): 27–57.

Rocke, Alan J. *Chemical Atomism in the Nineteenth Century: From Dalton to Cannizzaro*. Athens, Ohio: Ohio State University Press, 1984.

Rossi, Paolo. *The Birth of Modern Science*. Translated by Cynthia De Nardi Ipsen. Oxford: Wiley-Blackwell Publishers, 2001.

Ruthenberg, Klaus. "Paneth, Kant, and the Philosophy of Chemistry." *Foundations of Chemistry*, 11 (2009): 79–91.

Sadon-Goupil, M. *Révue de l' Éssai de statique chimique.* Paris: J. Vrin, 1980.

Sagal, Lynn. " On the Origin of Mitosing Cells." *Journal of Theoretical Biology* 14 (1967): 225–274.

Sander, Ian M., Julie L. Chaney, and Patricia L. Clark. "Expanding Anfinsen's Principle: Contributions of Synonymous Codon Selection to Rational Protein Design." *Journal of the American Chemical Society* 136 (2014): 858–861.

Sanger, F., and H. Tuppy. "The Amino Acid Sequence in the Phenylalanine Chain of Insulin. 2. The Investigation of Peptides from Enzymic Hydrolyzates." *Biochemical Journal* 9 (1951): 481–490.

Scerri, Eric. *Selected Papers on the Periodic Table.* London: Imperial College Press, 2009.

Scerri, Eric. *The Periodic Table: Its Story and Its Significance.* New York: Oxford University Press, 2019.

Schrödinger, Erwin. "Über das Verhältnis der Heisenberg-Born-Jordanschen Quantenmechanik zu der meinen." *Annalen der Physik* 79 (1926): 734–756.

Schrödinger, Erwin. "Quantization as a Problem of Proper Values." In *Erwin Schrödinger: Collected Papers on Wave Mechanics*, 1–40. London: Blackie & Son, 1928.

Schrödinger, Erwin. *What Is Life? The Physical Aspect of the Living Cell.* Cambridge: Cambridge University Press, 2012.

Science History Institute. "Julius Lothar Meyer and Dmitri Ivanovich Mendeleev." Available at: https://www.sciencehistory.org/historical-profile/julius-lothar-meyer-and-dmitri-ivanovich-mendeleev.

Sebeok, Thomas A. "Biosemiotics: Its Roots, Proliferation, and Prospects." *Semiotica* 134 (2001): 61–78.

Sennert, Daniel. *De chymicorum cum Aristotelicis et Galenicis consensu ac dissensu.* Wittenberg: Zacharia Shurerum, 1619.

Shapin, Steven, and Simon Schaffer. *Leviathan and the Air Pump: Hobbes, Boyle, and the Experimental Life.* Princeton, NJ: Princeton University Press, 2011.

Sommerhoff, Gerd. "Le caratteristiche astratte dei sistemi viventi." In *Teoria dei sistemi: Presupposti, caratteristiche e sviluppi del pensiero sistemico*, edited by F. E. Emery, 160–170. Milan: Franco Angeli, 2004.

Stahl, Georg Ernst. *Fundamenta Chymiae Dogmaticae & experimentalis.* Nuremberg: Wolfgang Moritz Endter, 1723.

Stone, G. B. "The Atomic View of Matter in the XVth, XVIth, and XVIIth Centuries." *Isis* 10, No. 2 (1928): 445–465.

Šubrt, Jiří. "Systemic Approach in Sociology: Reflections on Its Development, Current Status and Possibilities." *World Complexity Science Academy* 1, No. 9 (2020): 81–92. Available at: https://doi.org/10.46473/WCSAJ27240606/15-05-2020-0009//full/html.

Sytnik-Czetwertyński, Janusz. "The Philosophical Foundations of the Kinematic Atomism of Ruder Josip Boscovich." *Forum Philosophicum* 12 (2007): 139–155.

Thackray, Arnold. *Atoms and Powers: An Essay on Newtonian Matter-Theory and the Development of Chemistry.* Cambridge, MA: Harvard University Press, 1970.

Thomson, Joseph Joh. *The Atomic Theory.* Oxford: Clarendon Press, 1914.

Thurnbaugh, Peter J., Ruth E. Ley, Micah Hamady, Claire Fraser-Liggett, Rob Knight, and Jeffrey I. Gordon. "The Human Microbiome Project Exploring the Microbial Part of Ourselves in a Changing World." *Nature* 449 (2007): 804–810.

Tóth, Zoltán A. "Demonstration of Wöhler's Experiment: Preparation of Urea from Ammonium Chloride and Potassium Cyanate." *Journal of Chemical Education* 73 (1996): 539–540.

Truesdell, C. *Essays on the History of Mechanics.* Dordrecht: Springer, 1975.

Uversky, Vladimir, N. "Intrinsically Disordered Proteins and Their 'Mysterious' (Meta) Physics." *Frontiers in Physics* 7 (2019): 1–18. Available at: https://doi.org/10.3389/fphy.2019.00010.

Vihalemm, Rein. "Are Laws of Nature and Scientific Theories Peculiar in Chemistry? Scrutinizing Mendeleev's Discovery." *Foundations of Chemistry* 5 (2003): 7–22.

Villani, Giovanni. "Sostanze e reazioni chimiche: concetti di chimica teorica di interesse generale." *Epistemologia* 16 (1993): 191–212.

Villani, Giovanni. "Specificità della chimica." In *Philosophers in the Laboratory: Proceedings of the Meeting 'Riflessioni Epistemologiche e Metodologiche Sulla Chimica',* edited by Valeria Mosini, 163–180. Rome: MUSIS, 1996.

Villani, Giovanni. "Una *weltanschauung* scientifica: la chimica. In *Atti del Convegno triennale della SILFS: Prospettive della logica e della filosofia della scienza,* edited by Vito Michele Abrusci. Pisa: ETS, 1998.

Villani, Giovanni. *La chiave del mondo. Dalla filosofia alla scienza: l'onnipotenza delle molecole.* Naples: CUEN, 2001.

Villani, Giovanni. "Ruolo degli enti e dei processi in chimica e in fisica." *Epistemologia* 28 (2005): 199–218.

Villani, Giovanni. *Complesso e organizzato. Sistemi strutturati in fisica, chimica, biologia ed oltre.* Milan: Franco Angeli, 2008.

Villani, Giovanni. "La chimica: una scienza della complessità sistemica *ante litteram.*" In *Strutture di mondo. Il pensiero sistemico come specchio di una realtà complessa,* edited by Lucia Urbani Ulivi, 71–89. Bologna: il Mulino, 2010.

Villani, Giovanni. *Chemistry: A Systemic Complexity Science.* Pisa: Pisa University Press, 2017.

Villani, Giovanni, Elena Ghibaudi, and Luigi Cerruti. "The Orbital: A Pivotal Concept in the Relationship between Chemistry and Physics? A Comment to the Work by Fortin and Coauthors." *Foundations of Chemistry* 20 (2018): 89–97.

Virchow, Rud. *Vecchio e nuovo vitalismo.* Translated into the Italian by Vincenzo Cappelletti. Bari: Laterza, 1969.

Vries, Hugo de. *Intracellular Pangenesis: Including a Paper on Fertilization and Hybridization.* Chicago: Open Court, 1910.

Watson, James D., and Francis H. S. Crick. "Molecular Structure of Nucleic Acids: A Structure for Deoxyribose Nucleic Acid." *Nature* 171 (1953): 737–738.

Weinberg, Steven. "Newtonianism, Reductionism and the Art of Congressional Testimony." *Nature* 330 (1987): 433–443.

Whyte, Lancelot Law. "Boscovich's Atomism." In *Rujer Joseph Boscovich, 1711–1787: Studies of His Life and Work on the 250th Anniversary of His Birth,* edited by Lancelot Law Whyte. 102–126. London: Allen and Unwin, 1961.

Wilkins, M. H. F., A. R. Stokes, and H. R. Wilson. "Molecular Structure of Nucleic Acids: Molecular Structure of Deoxypentose Nucleic Acids." *Nature* 171 (1953): 738–740.

Wilson, Allan C. "Le basi molecolari dell'evoluzione." *Le Scienze* 208, No. 11 (1985): 166–175.

Wisniak, Jaime. "Claude-Louis Berthollet." *Revista CENIC. Ciencias Químicas* 39, No. 1 (2008): 45–55.

Wöhler, Friedrich. "Analytische Versuche Über Die Cyansäure." *Annalen der Physik* 77 (1824): 117–124.

Wöhler, Friedrich. "Correspondence from Wohler to Berzelius (February 22, 1828)." In *Briefwechsel zwischen J. Berzelius und F. Wohler*, edited by O. Wallach, 205–208. Leipzig: Engelmann, 1901.

Woese, Carl R. "A New Biology for a New Century." *Microbiology and Molecular Biology Review* 68 (2004): 173–186.

Wolfe, Charles T. "Endowed Molecules and Emergent Organization: The Maupertuis-Diderot Debate." *Early Science and Medicine* 15, No. 1/2 (2010): 38–85.

Wolley, Richard Guy. "Must a Molecule Have a Shape?" *Journal of the American Chemical Society* 100, No. 4 (1978): 1073–1078,

Index

For the benefit of digital users, indexed terms that span two pages (e.g., 52–53) may, on occasion, appear on only one of those pages.